Leopold Von Sacher

Venus In Furs

Leopold Von Sacher

Venus In Furs

ISBN/EAN: 9783337336929

Printed in Europe, USA, Canada, Australia, Japan

Cover: Foto ©berggeist007 / pixelio.de

More available books at **www.hansebooks.com**

VENUS
IN
FURS

LEOPOLD VON SACHER-MASOCH

Information about Project Gutenberg
The Legal Small Print

Venus in Furs
by Leopold von Sacher-Masoch

The Project Gutenberg EBook of Venus in Furs

by Leopold von Sacher-Masoch Translated by Fernanda Savage

Welcome To The World of Free Plain Vanilla Electronic Texts

eBooks Readable By Both Humans and By Computers, Since 1971

*****These eBooks Were Prepared By Thousands of Volunteers!*****

Title: Venus in Furs

Author: Leopold von Sacher-Masoch Translated by Fernanda Savage

Release Date: November, 2004 [EBook #6852] [Yes, we are more than one year ahead of schedule] [This file was first posted on February 2, 2003]

Edition: 10

Language: English

Character set encoding: ASCII

*** START OF THE PROJECT GUTENBERG EBOOK VENUS IN FURS ***

Produced by Avinash Kothare, Tom Allen, Tiffany Vergon, Charles Aldarondo, Charles Franks and the Online Distributed Proofreading Team.

VENUS IN FURS

Of this book, intended for private circulation, only 1225 copies have been printed, and type afterward distributed.

VENUS IN FURS

By

LEOPOLD VON SACHER-MASOCH

Translated from the German

By

FERNANDA SAVAGE

INTRODUCTION

Leopold von Sacher-Masoch was born in Lemberg, Austrian Galicia, on January 27, 1836. He studied jurisprudence at Prague and Graz, and in 1857 became a teacher at the latter university. He published several historical works, but soon gave up his academic career to devote himself wholly to literature. For a number of years he edited the international review, *Auf der Hohe*, at Leipzig, but later removed to Paris, for he was always strongly Francophile. His last years he spent at Lindheim in Hesse, Germany, where he died on March 9, 1895. In 1873 he married Aurora von Rumelin, who wrote a number of novels under the pseudonym of Wanda von Dunajew, which it is interesting to note is the name of the heroine of *Venus in Furs*. Her sensational memoirs which have been the cause of considerable controversy were published in 1906.

During his career as writer an endless number of works poured from Sacher-Masoch's pen. Many of these were works of ephemeral journalism, and some of them unfortunately pure sensationalism, for economic necessity forced him to turn his pen to unworthy ends.

There is, however, a residue among his works which has a distinct literary and even greater psychological value. His principal literary ambition was never completely fulfilled. It was a somewhat programmatic plan to give a picture of contemporary life in all its various aspects and interrelations under the general title of the *Heritage of Cain*. This idea was probably derived from Balzac's *Comedie Humaine*. The whole was to be divided into six subdivisions with the general titles *Love, Property, Money, The State, War*, and *Death*. Each of these divisions in its turn consisted of six novels, of which the last was intended to summarize the author's conclusions and to present his solution for the problems set in the others.

This extensive plan remained unachieved, and only the first two parts, *Love* and *Property*, were completed. Of the other sections only fragments remain. The present novel, *Venus in Furs*, forms the fifth in the series, *Love*.

The best of Sacher-Masoch's work is characterized by a swift narration and a graphic representation of character and scene and a rich humor. The latter has made many of his shorter stories dealing with his native Galicia little masterpieces of local color.

There is, however, another element in his work which has caused his name to become as eponym for an entire series of phenomena at one end of the psycho-sexual scale. This gives his productions a peculiar psychological value, though it cannot be denied also a morbid tinge that makes them often repellent. However, it is well to remember that nature is neither good nor bad, neither altruistic nor egoistic, and that it operates through the human psyche as well as through crystals and plants and animals with the same inexorable laws.

Sacher-Masoch was the poet of the anomaly now generally known as *masochism*. By this is meant the desire on the part of the individual affected of desiring himself completely and unconditionally subject to the will of a person of the opposite sex, and being treated by this person as by a master, to be humiliated, abused, and tormented, even to the verge of death. This motive is treated in all its innumerable variations. As a creative artist Sacher-Masoch was, of course, on the quest for the absolute, and sometimes, when impulses in the human being assume an abnormal or exaggerated form, there is just for a moment a flash that gives a glimpse of the thing in itself.

If any defense were needed for the publication of work like Sacher-Masoch's it is well to remember that artists are the historians of the human soul and one might recall the wise and tolerant Montaigne's essay *On the Duty of Historians* where he says, "One may cover over secret actions, but to be silent on what all the world knows, and things which have had effects which are public and of so much consequence is an inexcusable defect."

And the curious interrelation between cruelty and sex, again and again, creeps into literature. Sacher-Masoch has not created anything new in this. He has simply taken an ancient motive and developed it frankly and consciously, until, it seems, there is nothing further to say on the subject. To the violent attacks which his books met he replied in a polemical work, *Uber den Wert der Kritik*.

It would be interesting to trace the masochistic tendency as it occurs throughout literature, but no more can be done than just to allude to a few instances. The theme recurs continually in the *Confessions* of Jean Jacques Rousseau; it explains the character of the chevalier in Prevost's *Manon l'Escault*. Scenes of this nature are found in Zola's *Nana*, in Thomas Otway's *Venice Preserved*, in Albert Juhelle's *Les Pecheurs d'Hommes*, in Dostojevski. In disguised and unrecognized form it constitutes the undercurrent of much of the sentimental literature of the present day, though in most cases the authors as well as the readers are unaware of the pathological elements out of which their characters are built.

In all these strange and troubled waters of the human spirit one might wish for something of the serene and simple attitude of the ancient world. Laurent Tailhade has an admirable passage in his *Platres et Marbres*, which is well worth reproducing in this connection:

"Toutefois, les Hellenes, dans, leurs cites de lumiere, de douceur et d'harmonie, avaient une indulgence qu'on peut nommer scientifique pour les troubles amoureux de l'esprit. S'ils ne regardaient pas l'aliene comme en proie a la vistation d'un dieu (idee orientale et fataliste), du moins ils savaient que l'amour est une sorte d'envoutement, une folie ou se manifeste l'animosite des puissances cosmiques. Plus tard, le christianisme enveloppa les ames de tenebres. Ce fut la grande nuite. L'Eglise condamna tout ce qui lui parut neuf ou menacant pour les dogmes implacable ui reduisaient le monde en esclavage."

Among Sacher-Masoch's works, *Venus in Furs* is one of the most typical and outstanding. In spite of melodramatic elements and other literary faults, it is unquestionably a sincere work, written without any idea of titillating morbid fancies. One feels that in the hero many subjective elements have been incorporated, which are a disadvantage to the work from the point of view of literature, but on the other hand raise the book beyond the sphere of art, pure and simple, and make it one of those appalling human documents which belong, part to science and part to psychology. It is the confession of a deeply unhappy man who could not master his personal tragedy of existence, and so sought to unburden his soul in writing down the things he felt and experienced. The reader who will approach the book from this angle and who will honestly put aside moral prejudices and prepossessions will come away from the perusal of this book with a deeper understanding of this poor miserable soul of ours and a light will be cast into dark places that lie latent in all of us.

Sacher-Masoch's works have held an established position in European letters for something like half a century, and the author himself was made a chevalier of the Legion of Honor by the French Government in 1883, on the occasion of his literary jubilee. When several years ago cheap reprints were brought out on the Continent and attempts were made by various guardians of morality--they exist in all countries-- to have them suppressed, the judicial decisions were invariably against the plaintiff and in favor of the publisher. Are Americans children that they must be protected from books which any European school-boy can purchase whenever he wishes? However, such seems to be the case, and this translation, which has long been in preparation, consequently appears in a limited edition printed for subscribers only. In another connection Herbert Spencer once used these words: "The ultimate result of shielding men from the effects of folly, is to fill the world with fools." They have a very pointed application in the case of a work like *Venus in Furs*.

F. S.

Atlantic City April, 1921

VENUS IN FURS

"But the Almighty Lord hath struck him, and hath delivered him into the hands of a woman."

--The Vulgate, Judith, xvi. 7.

My company was charming.

Opposite me by the massive Renaissance fireplace sat Venus; she was not a casual woman of the half-world, who under this pseudonym wages war against the enemy sex, like Mademoiselle Cleopatra, but the real, true goddess of love.

She sat in an armchair and had kindled a crackling fire, whose reflection ran in red flames over her pale face with its white eyes, and from time to time over her feet when she sought to warm them.

Her head was wonderful in spite of the dead stony eyes; it was all I could see of her. She had wrapped her marble-like body in a huge fur, and rolled herself up trembling like a cat.

"I don't understand it," I exclaimed, "It isn't really cold any longer. For two weeks past we have had perfect spring weather. You must be nervous."

"Much obliged for your spring," she replied with a low stony voice, and immediately afterwards sneezed divinely, twice in succession. "I really can't stand it here much longer, and I am beginning to understand--"

"What, dear lady?"

"I am beginning to believe the unbelievable and to understand the un-
understandable. All of a sudden I understand the Germanic virtue of
woman, and German philosophy, and I am no longer surprised that you of
the North do not know how to love, haven't even an idea of what love is."

"But, madame," I replied flaring up, "I surely haven't given you any
reason."

"Oh, you--" The divinity sneezed for the third time, and shrugged her
shoulders with inimitable grace. "That's why I have always been nice to
you, and even come to see you now and then, although I catch a cold every
time, in spite of all my furs. Do you remember the first time we met?"

"How could I forget it," I said. "You wore your abundant hair in brown
curls, and you had brown eyes and a red mouth, but I recognized you
immediately by the outline of your face and its marble-like pallor--you
always wore a violet-blue velvet jacket edged with squirrel-skin."

"You were really in love with the costume, and awfully docile."

"You have taught me what love is. Your serene form of worship let me
forget two thousand years."

"And my faithfulness to you was without equal!"

"Well, as far as faithfulness goes--"

"Ungrateful!"

"I will not reproach you with anything. You are a divine woman, but
nevertheless a woman, and like every woman cruel in love."

"What you call cruel," the goddess of love replied eagerly, "is simply the
element of passion and of natural love, which is woman's nature and makes
her give herself where she loves, and makes her love everything, that
pleases her."

"Can there be any greater cruelty for a lover than the unfaithfulness of the woman he loves?"

"Indeed!" she replied. "We are faithful as long as we love, but you demand faithfulness of a woman without love, and the giving of herself without enjoyment. Who is cruel there--woman or man? You of the North in general take love too soberly and seriously. You talk of duties where there should be only a question of pleasure."

"That is why our emotions are honorable and virtuous, and our relations permanent."

"And yet a restless, always unsatisfied craving for the nudity of paganism," she interrupted, "but that love, which is the highest joy, which is divine simplicity itself, is not for you moderns, you children of reflection. It works only evil in you. *As soon as you wish to be natural, you become common.* To you nature seems something hostile; you have made devils out of the smiling gods of Greece, and out of me a demon. You can only exorcise and curse me, or slay yourselves in bacchantic madness before my altar. And if ever one of you has had the courage to kiss my red mouth, he makes a barefoot pilgrimage to Rome in penitential robes and expects flowers to grow from his withered staff, while under my feet roses, violets, and myrtles spring up every hour, but their fragrance does not agree with you. Stay among your northern fogs and Christian incense; let us pagans remain under the debris, beneath the lava; do not disinter us. Pompeii was not built for you, nor our villas, our baths, our temples. You do not require gods. We are chilled in your world."

The beautiful marble woman coughed, and drew the dark sables still closer about her shoulders.

"Much obliged for the classical lesson," I replied, "but you cannot deny, that man and woman are mortal enemies, in your serene sunlit world as well as in our foggy one. In love there is union into a single being for a short time only, capable of only one thought, one sensation, one will, in order to be then further disunited. And you know this better than I;

whichever of the two fails to subjugate will soon feel the feet of the other on his neck--"

"And as a rule the man that of the woman," cried Madame Venus with proud mockery, "which you know better than I."

"Of course, and that is why I don't have any illusions."

"You mean you are now my slave without illusions, and for that reason you shall feel the weight of my foot without mercy."

"Madame!"

"Don't you know me yet? Yes, I am *cruel*--since you take so much delight in that word-and am I not entitled to be so? Man is the one who desires, woman the one who is desired. This is woman's entire but decisive advantage. Through his passion nature has given man into woman's hands, and the woman who does not know how to make him her subject, her slave, her toy, and how to betray him with a smile in the end is not wise."

"Exactly your principles," I interrupted angrily.

"They are based on the experience of thousands of years," she replied ironically, while her white fingers played over the dark fur. "The more devoted a woman shows herself, the sooner the man sobers down and becomes domineering. The more cruelly she treats him and the more faithless she is, the worse she uses him, the more wantonly she plays with him, the less pity she shows him, by so much the more will she increase his desire, be loved, worshipped by him. So it has always been, since the time of Helen and Delilah, down to Catherine the Second and Lola Montez."

"I cannot deny," I said, "that nothing will attract a man more than the picture of a beautiful, passionate, cruel, and despotic woman who wantonly changes her favorites without scruple in accordance with her whim--"

"And in addition wears furs," exclaimed the divinity.

"What do you mean by that?"

"I know your predilection."

"Do you know," I interrupted, "that, since we last saw each other, you have grown very coquettish."

"In what way, may I ask?"

"In that there is no way of accentuating your white body to greater advantage than by these dark furs, and that--"

The divinity laughed.

"You are dreaming," she cried, "wake up!" and she clasped my arm with her marble-white hand. "Do wake up," she repeated raucously with the low register of her voice. I opened my eyes with difficulty.

I saw the hand which shook me, and suddenly it was brown as bronze; the voice was the thick alcoholic voice of my cossack servant who stood before me at his full height of nearly six feet.

"Do get up," continued the good fellow, "it is really disgraceful."

"What is disgraceful?"

"To fall asleep in your clothes and with a book besides." He snuffed the candles which had burned down, and picked up the volume which had fallen from my hand, "with a book by"--he looked at the title page-- "by Hegel. Besides it is high time you were starting for Mr. Severin's who is expecting us for tea."

"A curious dream," said Severin when I had finished. He supported his arms on his knees, resting his face in his delicate, finely veined hands, and fell to pondering.

I knew that he wouldn't move for a long time, hardly even breathe. This actually happened, but I didn't consider his behavior as in any way remarkable. I had been on terms of close friendship with him for nearly three years, and gotten used to his peculiarities. For it cannot be denied that he was peculiar, although he wasn't quite the dangerous madman that the neighborhood, or indeed the entire district of Kolomea, considered him to be. I found his personality not only interesting--and that is why many also regarded me a bit mad--but to a degree sympathetic. For a Galician nobleman and land-owner, and considering his age--he was hardly over thirty--he displayed surprising sobriety, a certain seriousness, even pedantry. He lived according to a minutely elaborated, half-philosophical, half- practical system, like clock-work; not this alone, but also by the thermometer, barometer, aerometer, hydrometer, Hippocrates, Hufeland, Plato, Kant, Knigge, and Lord Chesterfield. But at times he had violent attacks of sudden passion, and gave the impression of being about to run with his head right through a wall. At such times every one preferred to get out of his way.

While he remained silent, the fire sang in the chimney and the large venerable samovar sang; and the ancient chair in which I sat rocking to and fro smoking my cigar, and the cricket in the old walls sang too. I let my eyes glide over the curious apparatus, skeletons of animals, stuffed birds, globes, plaster-casts, with which his room was heaped full, until by chance my glance remained fixed on a picture which I had seen often enough before. But to-day, under the reflected red glow of the fire, it made an indescribable impression on me.

It was a large oil painting, done in the robust full-bodied manner of the Belgian school. Its subject was strange enough.

A beautiful woman with a radiant smile upon her face, with abundant hair tied into a classical knot, on which white powder lay like a soft hoarfrost, was resting on an ottoman, supported on her left arm. She was nude in her dark furs. Her right hand played with a lash, while her bare foot rested carelessly on a man, lying before her like a slave, like a dog. In the sharply outlined, but well-formed linaments of this man lay brooding melancholy and passionate devotion; he looked up to her with the ecstatic burning eye of a martyr. This man, the footstool for her feet, was Severin, but beardless, and, it seemed, some ten years younger.

"*Venus in Furs*," I cried, pointing to the picture. "That is the way I saw her in my dream."

"I, too," said Severin, "only I dreamed my dream with open eyes."

"Indeed?"

"It is a tiresome story."

"Your picture apparently suggested my dream," I continued. "But do tell me what it means. I can imagine that it played a role in your life, and perhaps a very decisive one. But the details I can only get from you."

"Look at its counterpart," replied my strange friend, without heeding my question.

The counterpart was an excellent copy of Titian's well-known "Venus with the Mirror" in the Dresden Gallery.

"And what is the significance?"

Severin rose and pointed with his finger at the fur with which Titian garbed his goddess of love.

"It, too, is a 'Venus in Furs,'" he said with a slight smile. "I don't believe that the old Venetian had any secondary intention. He simply painted the portrait of some aristocratic Mesalina, and was tactful enough to let Cupid hold the mirror in which she tests her majestic allure with cold satisfaction. He looks as though his task were becoming burdensome enough. The picture is painted flattery. Later an 'expert' in the Rococo period baptized the lady with the name of Venus. The furs of the despot in which Titian's fair model wrapped herself, probably more for fear of a cold than out of modesty, have become a symbol of the tyranny and cruelty that constitute woman's essence and her beauty.

"But enough of that. The picture, as it now exists, is a bitter satire on our love. Venus in this abstract North, in this icy Christian world, has to creep into huge black furs so as not to catch cold--"

Severin laughed, and lighted a fresh cigarette.

Just then the door opened and an attractive, stoutish, blonde girl entered. She had wise, kindly eyes, was dressed in black silk, and brought us cold meat and eggs with our tea. Severin took one of the latter, and decapitated it with his knife.

"Didn't I tell you that I want them soft-boiled?" he cried with a violence that made the young woman tremble.

"But my dear Sevtchu--" she said timidly.

"Sevtchu, nothing," he yelled, "you are to obey, obey, do you understand?" and he tore the *kantchuk* [Footnote: A long whip with a short handle.] which was hanging beside the weapons from its hook.

The woman fled from the chamber quickly and timidly like a doe.

"Just wait, I'll get you yet," he called after her.

"But Severin," I said placing my hand on his arm, "how can you treat a pretty young woman thus?"

"Look at the woman," he replied, blinking humorously with his eyes. "Had I flattered her, she would have cast the noose around my neck, but now, when I bring her up with the *kantchuk*, she adores me."

"Nonsense!"

"Nonsense, nothing, that is the way you have to break in women."

"Well, if you like it, live like a pasha in your harem, but don't lay down theories for me--"

"Why not," he said animatedly. "Goethe's 'you must be hammer or anvil' is absolutely appropriate to the relation between man and woman. Didn't Lady Venus in your dream prove that to you? Woman's power lies in man's passion, and she knows how to use it, if man doesn't understand himself. He has only one choice: to be the *tyrant* over or the *slave* of woman. As soon as he gives in, his neck is under the yoke, and the lash will soon fall upon him."

"Strange maxims!"

"Not maxims, but experiences," he replied, nodding his head, "*I have actually felt the lash*. I am cured. Do you care to know how?"

He rose, and got a small manuscript from his massive desk, and put it in front of me.

"You have already asked about the picture. I have long owed you an explanation. Here--read!"

Severin sat down by the chimney with his back toward me, and seemed to dream with open eyes. Silence had fallen again, and again the fire sang in the chimney, and the samovar and the cricket in the old walls. I opened the manuscript and read:

CONFESSIONS OF A SUPERSENSUAL MAN.

The margin of the manuscript bore as motto a variation of the well- known lines from *Faust*:

"Thou supersensual sensual woer A woman leads you by the nose."
--MEPHISTOPHELES.

I turned the title-page and read: "What follows has been compiled from my diary of that period, because it is impossible ever frankly to write of one's past, but in this way everything retains its fresh colors, the colors of the present."

Gogol, the Russian Moliere, says--where? well, somewhere--"the real comic muse is the one under whose laughing mask tears roll down."

A wonderful saying.

So I have a very curious feeling as I am writing all this down. The atmosphere seems filled with a stimulating fragrance of flowers, which overcomes me and gives me a headache. The smoke of the fireplace curls and condenses into figures, small gray-bearded kokolds that mockingly point their finger at me. Chubby-cheeked cupids ride on the arms of my chair and on my knees. I have to smile involuntarily, even laugh aloud, as I am writing down my adventures. Yet I am not writing with ordinary ink, but with red blood that drips from my heart. All its wounds long scarred over have opened and it throbs and hurts, and now and then a tear falls on the paper.

The days creep along sluggishly in the little Carpathian health- resort. You see no one, and no one sees you. It is boring enough to write idyls. I would have leisure here to supply a whole gallery of paintings, furnish a theater with new pieces for an entire season, a dozen virtuosos with concertos, trios, and duos, but--what am I saying--the upshot of it all is that I don't do much more than to stretch the canvas, smooth the bow, line the scores. For I am--no false modesty, Friend Severin; you can lie to others, but you don't quite succeed any longer in lying to yourself--I am nothing but a dilettante, a dilettante in painting, in poetry, in music, and several other of the so-called unprofitable arts, which, however, at present secure for their masters the income of a cabinet minister, or even that of a minor potentate. Above all else I am a dilettante in life.

Up to the present I have lived as I have painted and written poetry. I never got far beyond the preparation, the plan, the first act, the first stanza. There

are people like that who begin everything, and never finish anything. I am
such a one.

But what am I saying?

To the business in hand.

I lie in my window, and the miserable little town, which fills me with
despondency, really seems infinitely full of poetry. How wonderful the
outlook upon the blue wall of high mountains interwoven with golden
sunlight; mountain-torrents weave through them like ribbons of silver! How
clear and blue the heavens into which snowcapped crags project; how green
and fresh the forested slopes; the meadows on which small herds graze,
down to the yellow billows of grain where reapers stand and bend over and
rise up again.

The house in which I live stands in a sort of park, or forest, or wilderness,
whatever one wants to call it, and is very solitary.

Its sole inhabitants are myself, a widow from Lemberg, and Madame
Tartakovska, who runs the house, a little old woman, who grows older and
smaller each day. There are also an old dog that limps on one leg, and a
young cat that continually plays with a ball of yarn. This ball of yarn, I
believe, belongs to the widow.

She is said to be really beautiful, this widow, still very young, twenty-four
at the most, and very rich. She dwells in the first story, and I on the ground
floor. She always keeps the green blinds drawn, and has a balcony entirely
overgrown with green climbing- plants. I for my part down below have a
comfortable, intimate arbor of honeysuckle, in which I read and write and
paint and sing like a bird among the twigs. I can look up on the balcony.
Sometimes I actually do so, and then from time to time a white gown
gleams between the dense green network.

Really the beautiful woman up there doesn't interest me very much, for I
am in love with someone else, and terribly unhappy at that; far more

unhappy than the Knight of Toggenburg or the Chevalier in Manon l'Escault, because the object of my adoration is of stone.

In the garden, in the tiny wilderness, there is a graceful little meadow on which a couple of deer graze peacefully. On this meadow is a stone statue of Venus, the original of which, I believe, is in Florence. This Venus is the most beautiful woman I have ever seen in all my life.

That, however, does not signify much, for I have seen few beautiful women, or rather few women at all. In love too, I am a dilettante who never got beyond the preparation, the first act.

But why talk in superlatives, as if something that is beautiful could be surpassed?

It is sufficient to say that this Venus is beautiful. I love her passionately with a morbid intensity; madly as one can only love a woman who never responds to our love with anything but an eternally uniform, eternally calm, stony smile. I literally adore her.

I often lie reading under the leafy covering of a young birch when the sun broods over the forest. Often I visit that cold, cruel mistress of mine by night and lie on my knees before her, with the face pressed against the cold pedestal on which her feet rest, and my prayers go up to her.

The rising moon, which just now is waning, produces an indescribable effect. It seems to hover among the trees and submerges the meadow in its gleam of silver. The goddess stands as if transfigured, and seems to bathe in the soft moonlight.

Once when I was returning from my devotions by one of the walks leading to the house, I suddenly saw a woman's figure, white as stone, under the illumination of the moon and separated from me merely by a screen of trees. It seemed as if the beautiful woman of marble had taken pity on me, become alive, and followed me. I was seized by a nameless fear, my heart threatened to burst, and instead--

Well, I am a dilettante. As always, I broke down at the second stanza; rather, on the contrary, I did not break down, but ran away as fast as my legs would carry me.

* * * * *

What an accident! Through a Jew, dealing in photographs I secured a picture of my ideal. It is a small reproduction of Titian's "Venus with the Mirror." What a woman! I want to write a poem, but instead, I take the reproduction, and write on it: *Venus in Furs*.

You are cold, while you yourself fan flames. By all means wrap yourself in your despotic furs, there is no one to whom they are more appropriate, cruel goddess of love and of beauty!--After a while I add a few verses from Goethe, which I recently found in his paralipomena to *Faust*.

TO AMOR

"The pair of wings a fiction are, The arrows, they are naught but claws, The wreath conceals the little horns, For without any doubt he is Like all the gods of ancient Greece Only a devil in disguise."

Then I put the picture before me on my table, supporting it with a book, and looked at it.

I was enraptured and at the same time filled with a strange fear by the cold coquetry with which this magnificent woman draped her charms in her furs of dark sable; by the severity and hardness which lay in this cold marble-like face. Again I took my pen in hand, and wrote the following words:

"To love, to be loved, what happiness! And yet how the glamour of this pales in comparison with the tormenting bliss of worshipping a woman who makes a plaything out of us, of being the slave of a beautiful tyrant who treads us pitilessly underfoot. Even Samson, the hero, the giant, again put himself into the hands of Delilah, even after she had betrayed him, and again she betrayed him, and the Philistines bound him and put out his eyes which until the very end he kept fixed, drunken with rage and love, upon the beautiful betrayer."

I was breakfasting in my honey-suckle arbor, and reading in the Book of Judith. I envied the hero Holofernes because of the regal woman who cut off his head with a sword, and because of his beautiful sanguinary end.

"The almighty Lord hath struck him, and hath delivered him into the hands of a woman."

This sentence strangely impressed me.

How ungallant these Jews are, I thought. And their God might choose more becoming expressions when he speaks of the fair sex.

"The almighty Lord hath struck him, and hath delivered him into the hands of a woman," I repeated to myself. What shall I do, so that He may punish me?

Heaven preserve us! Here comes the housekeeper, who has again diminished somewhat in size overnight. And up there among the green twinings and garlandings the white gown gleams again. Is it Venus, or the widow?

This time it happens to be the widow, for Madame Tartakovska makes a courtesy, and asks me in her name for something to read. I run to my room, and gather together a couple of volumes.

Later I remember that my picture of Venus is in one of them, and now it and my effusions are in the hands of the white woman up there together. What will she say?

I hear her laugh.

Is she laughing at me?

It is full moon. It is already peering over the tops of the low hemlocks that fringe the park. A silvery exhalation fills the terrace, the groups of trees, all the landscape, as far as the eye can reach; in the distance it gradually fades away, like trembling waters.

I cannot resist. I feel a strange urge and call within me. I put on my clothes again and go out into the garden.

Some power draws me toward the meadow, toward her, who is my divinity and my beloved.

The night is cool. I feel a slight chill. The atmosphere is heavy with the odor of flowers and of the forest. It intoxicates.

What solemnity! What music round about! A nightingale sobs. The stars quiver very faintly in the pale-blue glamour. The meadow seems smooth, like a mirror, like a covering of ice on a pond.

The statue of Venus stands out august and luminous.

But--what has happened? From the marble shoulders of the goddess a large dark fur flows down to her heels. I stand dumbfounded and stare at her in amazement; again an indescribable fear seizes hold of me and I take flight.

I hasten my steps, and notice that I have missed the main path. As I am about to turn aside into one of the green walks I see Venus sitting before me on a stone bench, not the beautiful woman of marble, but the goddess of love herself with warm blood and throbbing pulses. She has actually come to life for me, like the statue that began to breathe for her creator. Indeed, the miracle is only half completed. Her white hair seems still to be of stone, and her white gown shimmers like moonlight, or is it satin? From her shoulders the dark fur flows. But her lips are already reddening and her cheeks begin to take color. Two diabolical green rays out of her eyes fall upon me, and now she laughs.

Her laughter is very mysterious, very--I don't know. It cannot be described, it takes my breath away. I flee further, and after every few steps I have to pause to take breath. The mocking laughter pursues me through the dark leafy paths, across light open spaces, through the thicket where only single moonbeams can pierce. I can no longer find my way, I wander about utterly confused, with cold drops of perspiration on the forehead.

Finally I stand still, and engage in a short monologue.

It runs--well--one is either very polite to one's self or very rude.

I say to myself:

"Donkey!"

This word exercises a remarkable effect, like a magic formula, which sets me free and makes me master of myself.

I am perfectly quiet in a moment.

With considerable pleasure I repeat: "Donkey!"

Now everything is perfectly clear and distinct before my eyes again. There is the fountain, there the alley of box-wood, there the house which I am slowly approaching.

Yet--suddenly the appearance is here again. Behind the green screen through which the moonlight gleams so that it seems embroidered with silver, I again see the white figure, the woman of stone whom I adore, whom I fear and flee.

With a couple of leaps I am within the house and catch my breath and reflect.

What am I really, a little dilettante or a great big donkey?

A sultry morning, the atmosphere is dead, heavily laden with odors, yet stimulating. Again I am sitting in my honey-suckle arbor, reading in the Odyssey about the beautiful witch who transformed her admirers into beasts. A wonderful picture of antique love.

There is a soft rustling in the twigs and blades and the pages of my book rustle and on the terrace likewise there is a rustling.

A woman's dress--

She is there--Venus--but without furs--No, this time it is merely the widow--and yet--Venus-oh, what a woman!

As she stands there in her light white morning gown, looking at me, her slight figure seems full of poetry and grace. She is neither large, nor small; her head is alluring, piquant--in the sense of the period of the French marquises--rather than formally beautiful. What enchantment and softness, what roguish charm play about her none too small mouth! Her skin is so infinitely delicate, that the blue veins show through everywhere; even through the muslin covering her arms and bosom. How abundant her red hair-it is red, not blonde or golden- yellow--how diabolically and yet tenderly it plays around her neck! Now her eyes meet mine like green lightnings--they are green, these eyes of hers, whose power is so indescribable--green, but as are precious stones, or deep unfathomable mountain lakes.

She observes my confusion, which has even made me discourteous, for I have remained seated and still have my cap on my head.

She smiles roguishly.

Finally I rise and bow to her. She comes closer, and bursts out into a loud, almost childlike laughter. I stammer, as only a little dilettante or great big donkey can do on such an occasion.

Thus our acquaintance began.

The divinity asks for my name, and mentions her own.

Her name is Wanda von Dunajew.

And she is actually my Venus.

"But madame, what put the idea into your head?"

"The little picture in one of your books--"

"I had forgotten about it."

"The curious notes on its back--"

"Why curious?"

She looked at me.

"I have always wanted to know a real dreamer some time--for the sake of the change--and you seem one of the maddest of the tribe."

"Dear lady--in fact--" Again I fell victim to an odious, asinine stammering, and in addition blushed in a way that might have been appropriate for a youngster of sixteen, but not for me, who was almost a full ten years older--

"You were afraid of me last night."

"Really--of course--but won't you sit down?"

She sat down, and enjoyed my embarrassment--for actually I was even more afraid of her now in the full light of day. A delightful expression of contempt hovered about her upper lip.

"You look at love, and especially woman," she began, "as something hostile, something against which you put up a defense, even if unsuccessfully. You feel that their power over you gives you a sensation of pleasurable torture, of pungent cruelty. This is a genuinely modern point of view."

"You don't share it?"

"I do not share it," she said quickly and decisively, shaking her head, so that her curls flew up like red flames.

"The ideal which I strive to realize in my life is the serene sensuousness of the Greeks--pleasure without pain. I do not believe in the kind of love which is preached by Christianity, by the moderns, by the knights of the spirit. Yes, look at me, I am worse than a heretic, I am a pagan.

'Doest thou imagine long the goddess of love took counsel When in Ida's grove she was pleased with the hero Achilles?'

"These lines from Goethe's *Roman Elegy* have always delighted me.

"In nature there is only the love of the heroic age, 'when gods and goddesses loved.' At that time 'desire followed the glance, enjoyment desire.' All else is factitious, affected, a lie. Christianity, whose cruel emblem, the cross, has always had for me an element of the monstrous, brought something alien and hostile into nature and its innocent instincts.

"The battle of the spirit with the senses is the gospel of modern man. I do not care to have a share in it."

"Yes, Mount Olympus would be the place for you, madame," I replied, "but we moderns can no longer support the antique serenity, least of all in love. The idea of sharing a woman, even if it were an Aspasia, with another revolts us. We are jealous as is our God. For example, we have made a term abuse out of the name of the glorious Phryne.

"We prefer one of Holbein's meagre, pallid virgins, which is wholly ours to an antique Venus, no matter how divinely beautiful she is, but who loves Anchises to-day, Paris to-morrow, Adonis the day after. And if nature triumphs in us so that we give our whole glowing, passionate devotion to such a woman, her serene joy of life appears to us as something demonic and cruel, and we read into our happiness a sin which we must expiate."

"So you too are one of those who rave about modern women, those miserable hysterical feminine creatures who don't appreciate a real man in their somnambulistic search for some dream-man and masculine ideal. Amid tears and convulsions they daily outrage their Christian duties; they cheat and are cheated; they always seek again and choose and reject; they are never happy, and never give happiness. They accuse fate instead of calmly confessing that they want to love and live as Helen and Aspasia lived. Nature admits of no permanence in the relation between man and woman."

"But, my dear lady--"

"Let me finish. It is only man's egoism which wants to keep woman like some buried treasure. All endeavors to introduce permanence in love, the most changeable thing in this changeable human existence, have gone shipwreck in spite of religious ceremonies, vows, and legalities. Can you deny that our Christian world has given itself over to corruption?"

"But--"

"But you are about to say, the individual who rebels against the arrangements of society is ostracized, branded, stoned. So be it. I am willing to take the risk; my principles are very pagan. I will live my own life as it pleases me. I am willing to do without your hypocritical respect; I prefer to be happy. The inventors of the Christian marriage have done well, simultaneously to invent immortality. I, however, have no wish to live eternally. When with my last breath everything as far as Wanda von Dunajew is concerned comes to an end here below, what does it profit me whether my pure spirit joins the choirs of angels, or whether my dust goes into the formation of new beings? Shall I belong to one man whom I don't love, merely because I have once loved him? No, I do not renounce; I love everyone who pleases me, and give happiness to everyone who loves me. Is that ugly? No, it is more beautiful by far, than if cruelly I enjoy the tortures, which my beauty excites, and virtuously reject the poor fellow who is pining away for me. I am young, rich, and beautiful, and I live serenely for the sake of pleasure and enjoyment."

While she was speaking her eyes sparkled roguishly, and I had taken hold of her hands without exactly knowing what to do with them, but being a genuine dilettante I hastily let go of them again.

"Your frankness," I said, "delights me, and not it alone--"

My confounded dilettantism again throttled me as though there were a rope around my neck.

"You were about to say--"

"I was about to say--I was--I am sorry--I interrupted you."

"How, so?"

A long pause. She is doubtless engaging in a monologue, which translated into my language would be comprised in the single word, "donkey."

"If I may ask," I finally began, "how did you arrive at these--these conclusions?"

"Quite simply, my father was an intelligent man. From my cradle onward I was surrounded by replicas of ancient art; at ten years of age I read *Gil Blas*, at twelve *La Pucelle*. Where others had Hop-o'-my-thumb, Bluebeard, Cinderella, as childhood friends, mine were Venus and Apollo, Hercules and Lackoon. My husband's personality was filled with serenity and sunlight. Not even the incurable illness which fell upon him soon after our marriage could long cloud his brow. On the very night of his death he took me in his arms, and during the many months when he lay dying in his wheel chair, he often said jokingly to me: 'Well, have you already picked out a lover?' I blushed with shame. 'Don't deceive me,' he added on one occasion, 'that would seem ugly to me, but pick out an attractive lover, or preferably several. You are a splendid woman, but still half a child, and you need toys.'

"I suppose, I hardly need tell you that during his life time I had no lover; but it was through him that I have become what I am, a woman of Greece."

"A goddess," I interrupted.

"Which one," she smiled.

"Venus."

She threatened me with her finger and knitted her brows. "Perhaps, even a 'Venus in Furs.' Watch out, I have a large, very large fur, with which I could cover you up entirely, and I have a mind to catch you in it as in a net."

"Do you believe," I said quickly, for an idea which seemed good, in spite of its conventionality and triteness, flashed into my head, "do you believe that your theories could be carried into execution at the present time, that Venus would be permitted to stray with impunity among our railroads and telegraphs in all her undraped beauty and serenity?"

"*Undraped*, of course not, but in furs," she replied smiling, "would you care to see mine?"

"And then--"

"What then?"

"Beautiful, free, serene, and happy human beings, such as the Greeks were, are only possible when it is permitted to have *slaves* who will perform the prosaic tasks of every day for them and above all else labor for them."

"Of course," she replied playfully, "an Olympian divinity, such as I am, requires a whole army of slaves. Beware of me!"

"Why?"

I myself was frightened at the hardiness with which I uttered this "why"; it did not startle her in the least.

She drew back her lips a little so that her small white teeth became visible, and then said lightly, as if she were discussing some trifling matter, "Do you want to be my slave?"

"There is no equality in love," I replied solemnly. "Whenever it is a matter of choice for me of ruling or being ruled, it seems much more satisfactory to me to be the slave of a beautiful woman. But where shall I find the woman who knows how to rule, calmly, full of self-confidence, even harshly, and not seek to gain her power by means of petty nagging?"

"Oh, that might not be so difficult."

"You think--"

"I--for instance--" she laughed and leaned far back--"I have a real talent for despotism--I also have the necessary furs--but last night you were really seriously afraid of me!"

"Quite seriously."

"And now?"

"Now, I am more afraid of you than ever!"

We are together every day, I and--Venus; we are together a great deal. We breakfast in my honey-suckle arbor, and have tea in her little sitting-room. I have an opportunity to unfold all my small, very small talents. Of what use would have been my study of all the various sciences, my playing at all the arts, if I were unable in the case of a pretty, little woman--

But this woman is by no means little; in fact she impresses me tremendously. I made a drawing of her to-day, and felt particularly clearly, how inappropriate the modern way of dressing is for a cameo- head like hers. The configuration of her face has little of the Roman, but much of the Greek.

Sometimes I should like to paint her as Psyche, and then again as Astarte. It depends upon the expression in her eyes, whether it is vaguely dreamy, or half-consuming, filled with tired desire. She, however, insists that it be a portrait-likeness.

I shall make her a present of furs.

How could I have any doubts? If not for her, for whom would princely furs be suitable?

* * * * *

I was with her yesterday evening, reading the *Roman Elegies* to her. Then I laid the book aside, and improvised something for her. She seemed pleased; rather more than that, she actually hung upon my words, and her bosom heaved.

Or was I mistaken?

The rain beat in melancholy fashion on the window-panes, the fire crackled in the fireplace in wintery comfort. I felt quite at home with her, and for a moment lost all my fear of this beautiful woman; I kissed her hand, and she permitted it.

Then I sat down at her feet and read a short poem I had written for her.

VENUS IN FURS.

"Place thy foot upon thy slave, Oh thou, half of hell, half of dreams;
Among the shadows, dark and grave, Thy extended body softly gleams."

And--so on. This time I really got beyond the first stanza. At her request I
gave her the poem in the evening, keeping no copy. And now as I am
writing this down in my diary I can only remember the first stanza.

I am filled with a very curious sensation. I don't believe that I am in love
with Wanda; I am sure that at our first meeting, I felt nothing of the
lightning-like flashes of passion. But I feel how her extraordinary, really
divine beauty is gradually winding magic snares about me. It isn't any
spiritual sympathy which is growing in me; it is a physical subjection,
coming on slowly, but for that reason more absolutely.

I suffer under it more and more each day, and she--she merely smiles.

* * * * *

Without any provocation she suddenly said to me to-day: "You interest me.
Most men are very commonplace, without verve or poetry. In you there is a
certain depth and capacity for enthusiasm and a deep seriousness, which
delight me. I might learn to love you."

After a short but severe shower we went out together to the meadow and
the statue of Venus. All about us the earth steamed; mists rose up toward
heaven like clouds of incense; a shattered rainbow still hovered in the air.
The trees were still shedding drops, but sparrows and finches were already
hopping from twig to twig. They are twittering gaily, as if very much
pleased at something. Everything is filled with a fresh fragrance. We cannot
cross the meadow for it is still wet. In the sunlight it looks like a small pool,
and the goddess of love seems to rise from the undulations of its mirror-like
surface. About her head a swarm of gnats is dancing, which, illuminated by
the sun, seem to hover above her like an aureole.

Wanda is enjoying the lovely scene. As all the benches along the walk are
still wet, she supports herself on my arm to rest a while. A soft weariness

permeates her whole being, her eyes are half closed; I feel the touch of her breath on my cheek.

How I managed to get up courage enough I really don't know, but I took hold of her hand, asking,

"Could you love me?"

"Why not," she replied, letting her calm, clear look rest upon me, but not for long.

A moment later I am kneeling before her, pressing my burning face against the fragrant muslin of her gown.

"But Severin--this isn't right," she cried.

But I take hold of her little foot, and press my lips upon it.

"You are getting worse and worse!" she cried. She tore herself free, and fled rapidly toward the house, the while her adorable slipper remained in my hand.

Is it an omen?

* * * * *

All day long I didn't dare to go near her. Toward evening as I was sitting in my arbor her gay red head peered suddenly through the greenery of her balcony. "Why don't you come up?" he called down impatiently.

I ran upstairs, and at the top lost courage again. I knocked very lightly. She didn't say come-in, but opened the door herself, and stood on the threshold.

"Where is my slipper?"

"It is--I have--I want," I stammered.

"Get it, and then we will have tea together, and chat."

When I returned, she was engaged in making tea. I ceremoniously placed the slipper on the table, and stood in the corner like a child awaiting punishment.

I noticed that her brows were slightly contracted, and there was an expression of hardness and dominance about her lips which delighted me.

All of a sudden she broke out laughing.

"So--you are really in love--with me?"

"Yes, and I suffer more from it than you can imagine?"

"You suffer?" she laughed again.

I was revolted, mortified, annihilated, but all this was quite useless.

"Why?" she continued, "I like you, with all my heart."

She gave me her hand, and looked at me in the friendliest fashion.

"And will you be my wife?"

Wanda looked at me--how did she look at me? I think first of all with surprise, and then with a tinge of irony.

"What has given you so much courage, all at once?"

"Courage?"

"Yes courage, to ask anyone to be your wife, and me in particular?" She lifted up the slipper. "Was it through a sudden friendship with this? But joking aside. Do you really wish to marry me?"

"Yes."

"Well, Severin, that is a serious matter. I believe, you love me, and I care for you too, and what is more important each of us finds the other interesting. There is no danger that we would soon get bored, but, you know, I am a fickle person, and just for that reason I take marriage seriously. If I assume obligations, I want to be able to meet them. But I am afraid--no--it would hurt you."

"Please be perfectly frank with me," I replied.

"Well then honestly, I don't believe I could love a man longer than-- " She inclined her head gracefully to one side and mused.

"A year."

"What do you imagine--a month perhaps."

"Not even me?"

"Oh you--perhaps two."

"Two months!" I exclaimed.

"Two months is very long."

"You go beyond antiquity, madame."

"You see, you cannot stand the truth."

Wanda walked across the room and leaned back against the fireplace, watching me and resting one of her arms on the mantelpiece.

"What shall I do with you?" she began anew.

"Whatever you wish," I replied with resignation, "whatever will give you
pleasure."

"How illogical!" she cried, "first you want to make me your wife, and then
you offer yourself to me as something to toy with."

"Wanda--I love you."

"Now we are back to the place where we started. You love me, and want to
make me your wife, but I don't want to enter into a new marriage, because I
doubt the permanence of both my and your feelings."

"But if I am willing to take the risk with you?" I replied.

"But it also depends on whether I am willing to risk it with you," she said
quietly. "I can easily imagine belonging to one man for my entire life, but
he would have to be a whole man, a man who would dominate me, who
would subjugate me by his inate strength, do you understand? And every
man--I know this very well--as soon as he falls in love becomes weak,
pliable, ridiculous. He puts himself into the woman's hands, kneels down
before her. The only man whom I could love permanently would be he
before whom I should have to kneel. I've gotten to like you so much,
however, that I'll try it with you."

I fell down at her feet.

"For heaven's sake, here you are kneeling already," she said mockingly.
"You are making a good beginning." When I had risen again she continued,
"I will give you a year's time to win me, to convince me that we are suited
to each other, that we might live together. If you succeed, I will become
your wife, and a wife, Severin, who will conscientiously and strictly
perform all her duties. During this year we will live as though we were
married--"

My blood rose to my head.

In her eyes too there was a sudden flame--

"We will live together," she continued, "share our daily life, so that we may find out whether we are really fitted for each other. *I grant you all the rights of a husband, of a lover, of a friend.* Are you satisfied?"

"I suppose, I'll have to be?"

"You don't have to."

"Well then, I want to--"

"Splendid. That is how a man speaks. Here is my hand."

* * * * *

For ten days I have been with her every hour, except at night. All the time I was allowed to look into her eyes, hold her hands, listen to what she said, accompany her wherever she went.

My love seems to me like a deep, bottomless abyss, into which I subside deeper and deeper. There is nothing now which could save me from it.

This afternoon we were resting on the meadow at the foot of the Venus-statue. I plucked flowers and tossed them into her lap; she wound them into wreaths with which we adorned our goddess.

Suddenly Wanda looked at me so strangely that my senses became confused and passion swept over my head like a conflagration. Losing command over myself, I threw my arms about her and clung to her lips, and she--she drew me close to her heaving breast.

"Are you angry?" I then asked her.

"I am never angry at anything that is natural--" she replied, "but *I* am afraid you suffer."

"Oh, I am suffering frightfully."

"Poor friend!" she brushed my disordered hair back from my fore- head. "I hope it isn't through any fault of mine."

"No--" I replied,--"and yet my love for you has become a sort of madness. The thought that I might lose you, perhaps actually lose you, torments me day and night."

"But you don't yet possess me," said Wanda, and again she looked at me with that vibrant, consuming expression, which had already once before carried me away. Then she rose, and with her small transparent hands placed a wreath of blue anemones upon the ringletted white head of Venus. Half against my will I threw my arm around her body.

"I can no longer live without you, oh wonderful woman," I said. "Believe me, believe only this once, that this time it is not a phrase, not a thing of dreams. I feel deep down in my innermost soul, that my life belongs inseparably with yours. If you leave me, I shall perish, go to pieces."

"That will hardly be necessary, for I love you," she took hold of my chin, "you foolish man!"

"But you will be mine only under conditions, while I belong to you unconditionally--"

"That isn't wise, Severin," she replied almost with a start. "Don't you know me yet, do you absolutely refuse to know me? I am good when I am treated seriously and reasonably, but when you abandon yourself too absolutely to me, I grow arrogant--"

"So be it, be arrogant, be despotic," I cried in the fulness of exaltation, "only be mine, mine forever." I lay at her feet, embracing her knees.

"Things will end badly, my friend," she said soberly, without moving.

"It shall never end," I cried excitedly, almost violently. "Only death shall part us. If you cannot be mine, all mine and for always, then *I want to be your slave*, serve you, suffer everything from you, if only you won't drive me away."

"Calm yourself," she said, bending down and kissing my forehead, "I am really very fond of you, but your way is not the way to win and hold me."

"I want to do everything, absolutely everything, that you want, only not to lose you," I cried, "only not that, I cannot bear the thought."

"Do get up."

I obeyed.

"You are a strange person," continued Wanda. "You wish to possess me at any price?"

"Yes, at any price."

"But of what value, for instance, would that be?"--She pondered; a lurking uncanny expression entered her eyes--"If I no longer loved you, if I belonged to another."

A shudder ran through me. I looked at her She stood firmly and confident before me, and her eyes disclosed a cold gleam.

"You see," she continued, "the very thought frightens you." A beautiful smile suddenly illuminated her face.

"I feel a perfect horror, when I imagine, that the woman I love and who has responded to my love could give herself to another regardless of me. But have I still a choice? If I love such a woman, even unto madness, shall I turn my back to her and lose everything for the sake of a bit of boastful strength; shall I send a bullet through my brains? I have two ideals of woman. If I cannot obtain the one that is noble and simple, the woman who will faithfully and truly share my life, well then I don't want anything half-way or lukewarm. Then I would rather be subject to a woman without virtue, fidelity, or pity. Such a woman in her magnificent selfishness is likewise an ideal. If I am not permitted to enjoy the happiness of love, fully and wholly, I want to taste its pains and torments to the very dregs; I want to be maltreated and betrayed by the woman I love, and the more cruelly the better. This too is a luxury."

"Have you lost your senses," cried Wanda.

"I love you with all my soul," I continued, "with all my senses, and your presence and personality are absolutely essential to me, if I am to go on living. Choose between my ideals. Do with me what you will, make of me your husband or your slave."

"Very well," said Wanda, contracting her small but strongly arched brows, "it seems to me it would be rather entertaining to have a man, who interests me and loves me, completely in my power; at least I shall not lack pastime. You were imprudent enough to leave the choice to me. Therefore I choose; I want you to be my slave, I shall make a plaything for myself out of you!"

"Oh, please do," I cried half-shuddering, half-enraptured. "If the foundation of marriage depends on equality and agreement, it is likewise true that the greatest passions rise out of opposites. We are such opposites, almost enemies. That is why my love is part hate, part fear. In such a relation only one can be hammer and the other anvil. I wish to be the anvil. I cannot be happy when I look down upon the woman I love. I want to adore a woman, and this I can only do when she is cruel towards me."

"But, Severin," replied Wanda, almost angrily, "do you believe me capable of maltreating a man who loves me as you do, and whom I love?"

"Why not, if I adore you the more on this account? *It is possible to love really only that which stands above us,* a woman, who through her beauty, temperament, intelligence, and strength of will subjugates us and becomes a despot over us."

"Then that which repels others, attracts you."

"Yes. That is the strange part of me."

"Perhaps, after all, there isn't anything so very unique or strange in all your passions, for who doesn't love beautiful furs? And everyone knows and feels how closely sexual love and cruelty are related."

"But in my case all these elements are raised to their highest degree," I replied.

"In other words, reason has little power over you, and you are by nature, soft, sensual, yielding."

"Were the martyrs also soft and sensual by nature?"

"The martyrs?"

"On the contrary, they were *supersensual men,* who found enjoyment in suffering. They sought out the most frightful tortures, even death itself, as others seek joy, and as they were, so am I--*supersensual.*"

"Have a care that in being such, you do not become a martyr to love, the *martyr of a woman.*"

We are sitting on Wanda's little balcony in the mellow fragrant summer night. A twofold roof is above us, first the green ceiling of climbing-plants, and then the vault of heaven sown with innumerable stars. The low wailing love-call of a cat rises from the park. I am sitting on footstool at the feet of my divinity, and am telling her of my childhood.

"And even then all these strange tendencies were distinctly marked in you?" asked Wanda.

"Of course, I can't remember a time when I didn't have them. Even in my cradle, so mother has told me, I was *supersensual.* I scorned the healthy breast of my nurse, and had to be brought up on goats' milk. As a little boy I was mysteriously shy before women, which really was only an expression of an inordinate interest in them. I was oppressed by the gray arches and half-darknesses of the church, and actually afraid of the glittering altars and images of the saints. Secretly, however, I sneaked as to a secret joy to a plaster-Venus which stood in my father's little library. I kneeled down before her, and to her I said the prayers I had been taught--the Paternoster, the Ave Maria, and the Credo.

"Once at night I left my bed to visit her. The sickle of the moon was my light and showed me the goddess in a pale-blue cold light. I prostrated myself before her and kissed her cold feet, as I had seen our peasants do when they kissed the feet of the dead Savior.

"An irresistible yearning seized me.

"I got up and embraced the beautiful cold body and kissed the cold lips. A deep shudder fell upon me and I fled, and later in a dream, it seemed to me, as if the goddess stood beside my bed, threatening me with up-raised arm.

"I was sent to school early and soon reached the gymnasium. I passionately grasped at everything which promised to make the world of antiquity accessible to me. Soon I was more familiar with the gods of Greece than with the religion of Jesus. I was with Paris when he gave the fateful apple to Venus, I saw Troy burn, and followed Ulysses on his wanderings. The prototypes of all that is beautiful sank deep into my soul, and consequently at the time when other boys are coarse and obscene, I displayed an insurmountable aversion to everything base, vulgar, unbeautiful.

"To me, the maturing youth, love for women seemed something especially base and unbeautiful, for it showed itself to me first in all its commonness. I avoided all contact with the fair sex; in short, I was supersensual to madness.

"When I was about fourteen my mother had a charming chamber-maid, young, attractive, with a figure just budding into womanhood. I was sitting one day studying my Tacitus and growing enthusiastic over the virtues of the ancient Teutons, while she was sweeping my room. Suddenly she stopped, bent down over me, in the meantime holding fast to the broom, and a pair of fresh, full, adorable lips touched mine. The kiss of the enamoured little cat ran through me like a shudder, but I raised up my *Germania*, like a shield against the temptress, and indignantly left the room."

Wanda broke out in loud laughter. "It would, indeed, be hard to find another man like you, but continue."

"There is another unforgetable incident belonging to that period," I continued my story. "Countess Sobol, a distant aunt of mine, was visiting my parents. She was a beautiful majestic woman with an attractive smile. I, however, hated her, for she was regarded by the family as a sort of Messalina. My behavior toward her was as rude, malicious, and awkward as possible.

"One day my parents drove to the capital of the district. My aunt determined to take advantage of their absence, and to exercise judgment over me. She entered unexpectedly in her fur-lined *kazabaika*, [Footnote: A woman's jacket.] followed by the cook, kitchen-maid, and the cat of a chamber-maid whom I had scorned. Without asking any questions, they seized me and bound me hand and foot, in spite of my violent resistance. Then my aunt, with an evil smile, rolled up her sleeve and began to whip me with a stout switch. She whipped so hard that the blood flowed, and that, at last, notwithstanding my heroic spirit, I cried and wept and begged for mercy. She then had me untied, but I had to get down on my knees and thank her for the punishment and kiss her hand.

"Now you understand the supersensual fool! Under the lash of a beautiful woman my senses first realized the meaning of woman. In her fur-jacket she seemed to me like a wrathful queen, and from then on my aunt became the most desirable woman on God's earth.

"My Cato-like austerity, my shyness before woman, was nothing but an excessive feeling for beauty. In my imagination sensuality became a sort of cult. I took an oath to myself that I would not squander its holy wealth upon any ordinary person, but I would reserve it for an ideal woman, if possible for the goddess of love herself.

"I went to the university at a very early age. It was in the capital where my aunt lived. My room looked at that time like Doctor Faustus's. Everything in it was in a wild confusion. There were huge closets stuffed full of books, which I bought for a song from a Jewish dealer on the Servanica; [Footnote: The street of the Jews in Lemberg.] there were globes, atlases, flasks, charts of the heavens, skeletons of animals, skulls, the busts of eminent men. It looked as though Mephistopheles might have stepped out from behind the huge green store as a wandering scholiast at any moment.

"I studied everything in a jumble without system, without selection: chemistry, alchemy, history, astronomy, philosophy, law, anatomy, and literature; I read Homer, Virgil, Ossian, Schiller, Goethe, Shakespeare, Cervantes, Voltaire, Moliere, the Koran, the Kosmos, Casanova's Memoirs. I grew more confused each day, more fantastical, more supersensual. All the time a beautiful ideal woman hovered in my imagination. Every so and so often she appeared before me like a vision among my leather-bound books and dead bones, lying on a bed of roses, surrounded by cupids. Sometimes she appeared gowned like the Olympians with the stern white face of the plaster Venus; sometimes in braids of a rich brown, blue-eyes, in my aunt's red velvet *kazabaika,* trimmed with ermine.

"One morning when she had again risen out of the golden mist of my imagination in all her smiling beauty, I went to see Countess Sobol, who received me in a friendly, even cordial manner. She gave me a kiss of welcome, which put all my senses in a turmoil. She was probably about forty years old, but like most well-preserved women of the world, still very attractive. She wore as always her fur-edged jacket. This time it was one of green velvet with brown marten. But nothing of the sternness which had so delighted me the other time was now discernable.

"On the contrary, there was so little of cruelty in her that without any more ado she let me adore her.

"Only too soon did she discover my supersensual folly and innocence, and it pleased her to make me happy. As for myself--I was as happy as a young god. What rapture for me to be allowed to lie before her on my knees, and to kiss her hands, those with which she had scourged me! What marvellous hands they were, of beautiful form, delicate, rounded, and white, with adorable dimples! I really was in love with her hands only. I played with them, let them submerge and emerge in the dark fur, held them against the light, and was unable to satiate my eyes with them."

Wanda involuntarily looked at her hand; I noticed it, and had to smile.

"From the way in which the supersensual predominated in me in those days you can see that I was in love only with the cruel lashes I received from my aunt; and about two years later when I paid court to a young actress only in the roles she played. Still later I became the admirer of a respectable woman. She acted the part of irreproachable virtue, only in the end to betray me with a rich Jew. You see, it is because I was betrayed, sold, by a woman who feigned the strictest principles and the highest ideals, that I hate that sort of poetical, sentimental virtue so intensely. Give me rather a woman who is honest enough to say to me: I am a Pompadour, a Lucretia Borgia, and I am ready to adore her."

Wanda rose and opened the window.

"You have a curious way of arousing one's imagination, stimulating all one's nerves, and making one's pulses beat faster. You put an aureole on vice, provided only if it is honest. Your ideal is a daring courtesan of genius. Oh, you are the kind of man who will corrupt a woman to her very last fiber."

* * * * *

In the middle of the night there was a knock at my window; I got up, opened it, and was startled. Without stood "Venus in Furs," just as she had appeared to me the first time.

"You have disturbed me with your stories; I have been tossing about in bed, and can't go to sleep," she said. "Now come and stay with me."

"In a moment."

As I entered Wanda was crouching by the fireplace where she had kindled a small fire.

"Autumn is coming," she began, "the nights are really quite cold already. I am afraid you may not like it, but I can't put off my furs until the room is sufficiently warm."

"Not like it--you are joking--you know--" I threw my arm around her, and kissed her.

"Of course, I know, but why this great fondness for furs?"

"I was born with it," I replied. "I already had it as a child. Furthermore furs have a stimulating effect on all highly organized natures. This is due both to general and natural laws. It is a physical stimulus which sets you tingling, and no one can wholly escape it. Science has recently shown a certain relationship between electricity and warmth; at any rate, their effects upon the human organism are related. The torrid zone produces more passionate characters, a heated atmosphere stimulation. Likewise with electricity. This is the reason why the presence of cats exercises such a magic influence upon highly-organized men of intellect. This is why these long-tailed Graces of the animal kingdom, these adorable, scintillating electric batteries have been the favorite animal of a Mahommed, Cardinal Richelieu, Crebillon, Rousseau, Wieland."

"A woman wearing furs, then," cried Wanda, "is nothing else than a large cat, an augmented electric battery?"

"Certainly," I replied. "That is my explanation of the symbolic meaning which fur has acquired as the attribute of power and beauty. Monarchs and the dominant higher nobility in former times used it in this sense for their costume, exclusively; great painters used it only for queenly beauty. The most beautiful frame, which Raphael could find for the divine forms of Fornarina and Titian for the roseate body of his beloved, was dark furs."

"Thanks for the learned discourse on love," said Wanda, "but you haven't told me everything. You associate something entirely individual with furs."

"Certainly," I cried. "I have repeatedly told you that suffering has a peculiar attraction for me. Nothing can intensify my passion more than tyranny, cruelty, and especially the faithlessness of a beautiful woman. And I cannot imagine this woman, this strange ideal derived from an aesthetics of ugliness, this soul of Nero in the body of a Phryne, except in furs."

"I understand," Wanda interrupted. "It gives a dominant and imposing quality to a woman."

"Not only that," I continued. "You know I am *supersensual*. With me everything has its roots in the imagination, and thence it receives its nourishment. I was already pre-maturely developed and highly sensitive, when at about the age of ten the legends of the martyrs fell into my hands. I remember reading with a kind of horror, which really was rapture, of how they pined in prisons, were laid on the gridiron, pierced with arrows, boiled in pitch, thrown to wild animals, nailed to the cross, and suffered the most horrible torment with a kind of joy. To suffer and endure cruel torture from then on seemed to me exquisite delight, especially when it was inflicted by a beautiful woman, for ever since I can remember all poetry and everything demonic was for me concentrated in woman. I literally carried the idea into a sort of cult.

"I felt there was something sacred in sex; in fact, it was the only sacred thing. In woman and her beauty I saw something divine, because the most important function of existence--the continuation of the species--is her vocation. To me woman represented a personification of nature, *Isis*, and man was her priest, her slave. In contrast to him she was cruel like nature herself who tosses aside whatever has served her purposes as soon as she no longer has need for it. To him her cruelties, even death itself, still were sensual raptures.

"I envied King Gunther whom the mighty Brunhilde fettered on the bridal night, and the poor troubadour whom his capricious mistress had sewed in the skins of wolves to have him hunted like game. I envied the Knight Ctirad whom the daring Amazon Scharka craftily ensnared in a forest near Prague, and carried to her castle Divin, where, after having amused herself a while with him, she had him broken on the wheel--"

"Disgusting," cried Wanda. "I almost wish you might fall into the hands of a woman of their savage race. In the wolf's skin, under the teeth of the dogs, or upon the wheel, you would lose the taste for your kind of poetry."

"Do you think so? I hardly do."

"Have you actually lost your senses."

"Possibly. But let me go on. I developed a perfect passion for reading stories in which the extremest cruelties were described. I loved especially to look at pictures and prints which represented them. All the sanguinary tyrants that ever occupied a throne; the inquisitors who had the heretics tortured, roasted, and butchered; all the woman whom the pages of history have recorded as lustful, beautiful, and violent women like Libussa, Lucretia Borgia, Agnes of Hungary, Queen Margot, Isabeau, the Sultana Roxolane, the Russian Czarinas of last century--all these I saw in furs or in robes bordered with ermine."

"And so furs now rouse strange imaginings in you," said Wanda, and simultaneously she began to drape her magnificent fur-cloak coquettishly about her, so that the dark shining sable played beautifully around her bust and arms. "Well, how do you feel now, half broken on the wheel?"

Her piercing green eyes rested on me with a peculiar mocking satisfaction. Overcome by desire, I flung myself down before her, and threw my arms about her.

"Yes--you have awakened my dearest dream," I cried. "It has slept long enough."

"And this is?" She put her hand on my neck.

I was seized with a sweet intoxication under the influence of this warm little hand and of her regard, which, tenderly searching, fell upon me through her half-closed lids.

"To be the slave of a woman, a beautiful woman, whom I love, whom I worship."

"And who on that account maltreats you," interrupted Wanda, laughing.

"Yes, who fetters me and whips me, treads me underfoot, the while she gives herself to another."

"And who in her wantonness will go so far as to make a present of you to your successful rival when driven insane by jealousy you must meet him face to face, who will turn you over to his absolute mercy. Why not? This final tableau doesn't please you so well?"

I looked at Wanda frightened.

"You surpass my dreams."

"Yes, we women are inventive," she said, "take heed, when you find your ideal, it might easily happen, that she will treat you more cruelly than you anticipate."

"I am afraid that I have already found my ideal!" I exclaimed, burying my burning face in her lap.

"Not I?" exclaimed Wanda, throwing off her furs and moving about the room laughing. She was still laughing as I went downstairs, and when I stood musing in the yard, I still heard her peals of laughter above.

* * * * *

"Do you really then expect me to embody your ideal?" Wanda asked archly, when we met in the park to-day.

At first I could find no answer. The most antagonistic emotions were battling within me. In the meantime she sat down on one of the stone-benches, and played with a flower.

"Well--am I?"

I kneeled down and seized her hands.

"Once more I beg you to become my wife, my true and loyal wife; if you can't do that then become the embodiment of my ideal, absolutely, without reservation, without softness."

"You know I am ready at the end of a year to give you my hand, if you prove to be the man I am seeking," Wanda replied very seriously, "but I think you would be more grateful to me if through me you realized your imaginings. Well, which do you prefer?"

"I believe that everything my imagination has dreamed lies latent in your personality."

"You are mistaken."

"I believe," I continued, "that you enjoy having a man wholly in your power, torturing him--"

"No, no," she exclaimed quickly, "or perhaps--." She pondered.

"I don't understand myself any longer," she continued, "but I have a confession to make to you. You have corrupted my imagination and inflamed my blood. I am beginning to like the things you speak of. The enthusiasm with which you speak of a Pompadour, a Catherine the Second, and all the other selfish, frivolous, cruel women, carries me away and takes hold of my soul. It urges me on to become like those women, who in spite of their vileness were slavishly adored during their lifetime and still exert a miraculous power from their graves.

"You will end by making of me a despot in miniature, a domestic Pompadour."

"Well then," I said in agitation, "if all this is inherent in you, give way to this trend of your nature. Nothing half-way. If you can't be a true and loyal wife to me, be a demon."

I was nervous from loss of sleep, and the proximity of the beautiful woman affected me like a fever. I no longer recall what I said, but I remember that I kissed her feet, and finally raised her foot and put my neck under it. She withdrew it quickly, and rose almost angrily.

"If you love me, Severin," she said quickly, and her voice sounded sharp and commanding, "never speak to me of those things again. Understand, never! Otherwise I might really--" She smiled and sat down again.

"I am entirely serious," I exclaimed, half-raving. "I adore you so infinitely that I am willing to suffer anything from you, for the sake of spending my whole life near you."

"Severin, once more I warn you."

"Your warning is vain. Do with me what you will, as long as you don't drive me away."

"Severin," replied Wanda, "I am a frivolous young woman; it is dangerous for you to put yourself so completely in my power. You will end by actually becoming a plaything to me. Who will give warrant that I shall not abuse your insane desire?"

"Your own nobility of character."

"Power makes people over-bearing."

"Be it," I cried, "tread me underfoot."

Wanda threw her arms around my neck, looked into my eyes, and shook her head.

"I am afraid I can't, but I will try, for your sake, for I love you Severin, as I have loved no other man."

* * * * *

To-day she suddenly took her hat and shawl, and I had to go shopping with her. She looked at whips, long whips with a short handle, the kind that are used on dogs.

"Are these satisfactory?" said the shopkeeper.

"No, they are much too small," replied Wanda, with a side-glance at me. "I need a large--"

"For a bull-dog, I suppose?" opined the merchant.

"Yes," she exclaimed, "of the kind that are used in Russia for intractable slaves."

She looked further and finally selected a whip, at whose sight I felt a strange creeping sensation.

"Now good-by, Severin," she said. "I have some other purchases to make, but you can't go along."

I left her and took a walk. On the way back I saw Wanda coming out at a furrier's. She beckoned me.

"Consider it well," she began in good spirits, "I have never made a secret of how deeply your serious, dreamy character has fascinated me. The idea of seeing this serious man wholly in my power, actually lying enraptured at my feet, of course, stimulates me--but will this attraction last? Woman loves a man; she maltreats a slave, and ends by kicking him aside."

"Very well then, kick me aside," I replied, "when you are tired of me. I want to be your slave."

"Dangerous forces lie within me," said Wanda, after we had gone a few steps further. "You awaken them, and not to your advantage. You know how to paint pleasure, cruelty, arrogance in glowing colors. What would you say should I try my hand at them, and make you the first object of my experiments. I would be like Dionysius who had the inventor of the iron ox roasted within it in order to see whether his wails and groans really resembled the bellowing of an ox.

"Perhaps I am a female Dionysius?"

"Be it," I exclaimed, "and my dreams will be fulfilled. I am yours for good or evil, choose. The destiny that lies concealed within my breast drives me on--demoniacally--relentlessly."

"My Beloved,

I do not care to see you to-day or to-morrow, and not until evening the day after tomorrow, and then *as my slave*.

Your mistress

Wanda."

"As my slave" was underlined. I read the note which I received early in the morning a second time. Then I had a donkey saddled, an animal symbolic of learned professors, and rode into the mountains. I wanted to numb my desire, my yearning, with the magnificent scenery of the Carpathians. I am back, tired, hungry, thirsty, and more in love than ever. I quickly change my clothes, and a few moments later knock at her door.

"Come in!"

I enter. She is standing in the center of the room, dressed in a gown of white satin which floods down her body like light. Over it she wears a scarlet *kazabaika*, richly edged with ermine. Upon her powdered, snowy hair is a little diadem of diamonds. She stands with her arms folded across her breast, and with her brows contracted.

"Wanda!" I run toward her, and am about to throw my arm about her to kiss her. She retreats a step, measuring me from top to bottom.

"Slave!"

"Mistress!" I kneel down, and kiss the hem of her garment.

"That is as it should be."

"Oh, how beautiful you are."

"Do I please you?" She stepped before the mirror, and looked at herself with proud satisfaction.

"I shall become mad!"

Her lower lip twitched derisively, and she looked at me mockingly from behind half-closed lids.

"Give me the whip."

I looked about the room.

"No," she exclaimed, "stay as you are, kneeling." She went over to the fire-place, took the whip from the mantle-piece, and, watching me with a smile, let it hiss through the air; then she slowly rolled up the sleeve of her fur-jacket.

"Marvellous woman!" I exclaimed.

"Silence, slave!" She suddenly scowled, looked savage, and struck me with the whip. A moment later she threw her arm tenderly about me, and pityingly bent down to me. "Did I hurt you?" she asked, half- shyly, half-timidly.

"No," I replied, "and even if you had, pains that come through you are a joy. Strike again, if it gives you pleasure."

"But it doesn't give me pleasure."

Again I was seized with that strange intoxication.

"Whip me," I begged, "whip me without mercy."

Wanda swung the whip, and hit me twice. "Are you satisfied now?"

"No."

"Seriously, no?"

"Whip me, I beg you, it is a joy to me."

"Yes, because you know very well that it isn't serious," she replied, "because I haven't the heart to hurt you. This brutal game goes against my grain. Were I really the woman who beats her slaves you would be horrified."

"No, Wanda," I replied, "I love you more than myself; I am devoted to you for death and life. In all seriousness, you can do with me whatever you will, whatever your caprice suggests."

"Severin!"

"Tread me underfoot!" I exclaimed, and flung myself face to the floor before her.

"I hate all this play-acting," said Wanda impatiently.

"Well, then maltreat me seriously."

An uncanny pause.

"Severin, I warn you for the last time," began Wanda.

"If you love me, be cruel towards me," I pleaded with upraised eyes.

"If I love you," repeated Wanda. "Very well!" She stepped back and looked at me with a sombre smile. *Be then my slave, and know what it means to be delivered into the hands of a woman.* And at the same moment she gave me a kick.

"How do you like that, slave?"

Then she flourished the whip.

"Get up!"

I was about to rise.

"Not that way," she commanded, "on your knees."

I obeyed, and she began to apply the lash.

The blows fell rapidly and powerfully on my back and arms. Each one cut into my flesh and burned there, but the pains enraptured me. They came from her whom I adored, and for whom I was ready at any hour to lay down my life.

She stopped. "I am beginning to enjoy it," she said, "but enough for to-day. I am beginning to feel a demonic curiosity to see how far your strength goes. I take a cruel joy in seeing you tremble and writhe beneath my whip, and in hearing your groans and wails; I want to go on whipping without pity until you beg for mercy, until you lose your senses. You have awakened dangerous elements in my being. But now get up."

I seized her hand to press it to my lips.

"What impudence."

She shoved me away with her foot.

"Out of my sight, slave!"

* * * * *

After having spent a feverish night filled with confused dreams, I awoke. Dawn was just beginning to break.

How much of what was hovering in my memory was true; what had I actually experienced and what had I dreamed? That I had been whipped was certain. I can still feel each blow, and count the burning red stripes on my body. And *she* whipped me. Now I know everything.

My dream has become truth. How does it make me feel? Am I disappointed in the realization of my dream?

No, I am merely somewhat tired, but her cruelty has enraptured me. Oh, how I love her, adore her! All this cannot express in the remotest way my feeling for her, my complete devotion to her. What happiness to be her slave!

* * * * *

She calls to me from her balcony. I hurry upstairs. She is standing on the threshold, holding out her hand in friendly fashion. "I am ashamed of myself," she says, while I embrace her, and she hides her head against my breast.

"Why?"

"Please try to forget the ugly scene of yesterday," she said with quivering voice, "I have fulfilled your mad wish, now let us be reasonable and happy and love each other, and in a year I will be your wife."

"My mistress," I exclaimed, "and I your slave!"

"Not another word of slavery, cruelty, or the whip," interrupted Wanda. "I shall not grant you any of those favors, none except wearing my fur-jacket; come and help me into it."

* * * * *

The little bronze clock on which stood a cupid who had just shot his bolt struck midnight.

I rose, and wanted to leave.

Wanda said nothing, but embraced me and drew me back on the ottoman. She began to kiss me anew, and this silent language was so comprehensible, so convincing--

And it told me more than I dared to understand.

A languid abandonment pervaded Wanda's entire being. What a voluptuous softness there was in the gloaming of her half-closed eyes, in the red flood of her hair which shimmered faintly under the white powder, in the red and white satin which crackled about her with every movement, in the swelling ermine of the *kazabaika* in which she carelessly nestled.

"Please," I stammered, "but you will be angry with me."

"Do with me what you will," she whispered.

"Well, then whip me, or I shall go mad."

"Haven't I forbidden you," said Wanda sternly, "but you are incorrigible."

"Oh, I am so terribly in love." I had sunken on my knees, and was burying my glowing face in her lap.

"I really believe," said Wanda thoughtfully, "that your madness is nothing but a demonic, unsatisfied sensuality. *Our unnatural way of life must generate such illnesses.* Were you less virtuous, you would be completely sane."

"Well then, make me sane," I murmured. My hands were running through her hair and playing tremblingly with the gleaming fur, which rose and fell like a moonlit wave upon her heaving bosom, and drove all my senses into confusion.

And I kissed her. No, she kissed me savagely, pitilessly, as if she wanted to slay me with her kisses. I was as in a delirium, and had long since lost my reason, but now I, too, was breathless. I sought to free myself.

"What is the matter?" asked Wanda.

"I am suffering agonies."

"You are suffering--" she broke out into a loud amused laughter.

"You laugh!" I moaned, "have you no idea--"

She was serious all of a sudden. She raised my head in her hands, and with
a violent gesture drew me to her breast.

"Wanda," I stammered.

"Of course, you enjoy suffering," she said, and laughed again, "but wait, I'll
bring you to your senses."

"No, I will no longer ask," I exclaimed, "whether you want to belong to me
for always or for only a brief moment of intoxication. I want to drain my
happiness to the full. You are mine now, and I would rather lose you than
never to have had you."

"Now you are sensible," she said. She kissed me again with her murderous
lips. I tore the ermine apart and the covering of lace and her naked breast
surged against mine.

Then my senses left me--

The first thing I remember is the moment when I saw blood dripping from
my hand, and she asked apathetically: "Did you scratch me?"

"No, I believe, I have bitten you."

* * * * *

It is strange how every relation in life assumes a different face as soon as a
new person enters.

We spent marvellous days together; we visited the mountains and lakes, we
read together, and I completed Wanda's portrait. And how we loved one
another, how beautiful her smiling face was!

Then a friend of hers arrived, a divorced woman somewhat older, more
experienced, and less scrupulous than Wanda. Her influence is already
making itself felt in every direction.

Wanda wrinkles her brows, and displays a certain impatience with me.

Has she ceased loving me?

* * * * *

For almost a fortnight this unbearable restraint has lain upon us. Her friend lives with her, and we are never alone. A circle of men surrounds the young women. With my seriousness and melancholy I am playing an absurd role as lover. Wanda treats me like a stranger.

To-day, while out walking, she staid behind with me. I saw that this was done intentionally, and I rejoiced. But what did she tell me?

"My friend doesn't understand how I can love you. She doesn't think you either handsome or particularly attractive otherwise. She is telling me from morning till night about the glamour of the frivolous life in the capital, hinting at the advantages to which I could lay claim, the large parties which I would find there, and the distinguished and handsome admirers which I would attract. But of what use is all this, since it happens that I love you."

For a moment I lost my breath, then I said: "I have no wish to stand in the way of your happiness, Wanda. Do not consider me." Then I raised my hat, and let her go ahead. She looked at me surprised, but did not answer a syllable.

When by chance I happened to be close to her on the way back, she secretly pressed my hand. Her glance was so radiant, so full of promised happiness, that in a moment all the torments of these days were forgotten and all their wounds healed.

I now am aware again of how much I love her.

* * * * *

"My friend has complained about you," said Wanda to-day.

"Perhaps she feels that I despise her."

"But why do you despise her, you foolish young man?" exclaimed Wanda, pulling my ears with both hands.

"Because she is a hypocrite," I said. "I respect only a woman who is actually virtuous, or who openly lives for pleasure's sake."

"Like me, for instance," replied Wanda jestingly, "but you see, child, a woman can only do that in the rarest cases. She can neither be as gaily sensual, nor as spiritually free as man; her state is always a mixture of the sensual and spiritual. Her heart desires to enchain man permanently, while she herself is ever subject to the desire for change. The result is a conflict, and thus usually against her wishes lies and deception enter into her actions and personality and corrupt her character."

"Certainly that is true," I said. "The transcendental character with which woman wants to stamp love leads her to deception."

"But the world likewise demands it," Wanda interrupted. "Look at this woman. She has a husband and a lover in Lemberg and has found a new admirer here. She deceives all three and yet is honored by all and respected by the world."

"I don't care," I exclaimed, "but she is to leave you alone; she treats you like an article of commerce."

"Why not?" the beautiful woman interrupted vivaciously. "Every woman has the instinct or desire to draw advantage out of her attractions, and much is to be said for giving one's self without love or pleasure because if you do it in cold blood, you can reap profit to best advantage."

"Wanda, what are you saying?"

"Why not?" she said, "and take note of what I am about to say to you. *Never feel secure with the woman you love,* for there are more dangers in woman's nature than you imagine. Women are neither as *good* as their admirers and defenders maintain, nor as *bad* as their enemies make them out to be. *Woman's character is characterlessness.* The best woman will momentarily go down into the mire, and the worst unexpectedly rises to deeds of greatness and goodness and puts to shame those that despise her. No woman is so good or so bad, but that at any moment she is capable of the most diabolical as well as of the most divine, of the filthiest as well as of the purest, thoughts, emotions, and actions. In spite of all the advances of civilization, woman has remained as she came out of the hand of nature. She has the nature of a savage, who is faithful or faithless, magnanimous or cruel, according to the impulse that dominates at the moment. Throughout history it has always been a serious deep culture which has produced moral character. Man even when he is selfish or evil always follows *principles,* woman never follows anything but *impulses.* Don't ever forget that, and never feel secure with the woman you love."

* * * * *

Her friend has left. At last an evening alone with her again. It seems as if Wanda had saved up all the love, which had been kept from her, for this superlative evening; never had she been so kind, so near, so full of tenderness.

What happiness to cling to her lips, and to die away in her arms! In a state of relaxation and wholly mine, her head rests against my breast, and with drunken rapture our eyes seek each other.

I cannot yet believe, comprehend, that this woman is mine, wholly mine.

"She is right on one point," Wanda began, without moving, without opening her eyes, as if she were asleep.

"Who?"

She remained silent.

"Your friend?"

She nodded. "Yes, she is right, you are not a man, you are a dreamer, a charming cavalier, and you certainly would be a priceless slave, but I cannot imagine you as husband."

I was frightened.

"What is the matter? You are trembling?"

"I tremble at the thought of how easily I might lose you," I replied.

"Are you made less happy now, because of this?" she replied. "Does it rob you of any of your joys, that I have belonged to another before I did to you, that others after you will possess me, and would you enjoy less if another were made happy simultaneously with you?"

"Wanda!"

"You see," she continued, "that would be a way out. You won't ever lose me then. I care deeply for you and intellectually we are harmonious, and I should like to live with you always, if in addition to you I might have--"

"What an idea," I cried. "You fill me with a sort of horror."

"Do you love me any the less?"

"On the contrary."

Wanda had raised herself on her left arm. "I believe," she said, "that to hold a man permanently, it is vitally important not to be faithful to him. What honest woman has ever been as devotedly loved as a hetaira?"

"There is a painful stimulus in the unfaithfulness of a beloved woman. It is the highest kind of ecstacy."

"For you, too?" Wanda asked quickly.

"For me, too."

"And if I should give you that pleasure," Wanda exclaimed mockingly.

"I shall suffer terrible agonies, but I shall adore you the more," I replied. "But you would never deceive me, you would have the daemonic greatness of saying to me: I shall love no one but you, but I shall make happy whoever pleases me."

Wanda shook her head. "I don't like deception, I am honest, but what man exists who can support the burden of truth. Were I say to you: this serene, sensual life, this paganism is my ideal, would you be strong enough to bear it?"

"Certainly. I could endure anything so as not to lose you. I feel how little I really mean to you."

"But Severin--"

"But it is so," said I, "and just for that reason--"

"For that reason you would--" she smiled roguishly--"have I guessed it?"

"Be your slave!" I exclaimed. "Be your unrestricted property, without a will of my own, of which you could dispose as you wished, and which would therefore never be a burden to you. While you drink life at its fullness, while surrounded by luxury, you enjoy the serene happiness and Olympian love, I want to be your servant, put on and take off your shoes."

"You really aren't so far from wrong," replied Wanda, "for only as my slave could you endure my loving others. Furthermore the freedom of enjoyment of the ancient world is unthinkable without slavery. It must give one a feeling of like unto a god to see a man kneel before one and tremble. I want a slave, do you hear, Severin?"

"Am I not your slave?"

"Then listen to me," said Wanda excitedly, seizing my hand. "I want to be yours, as long as I love you."

"A month?"

"Perhaps, even two."

"And then?"

"Then you become my slave."

"And you?"

"I? Why do you ask? I am a goddess and sometimes I descend from my Olympian heights to you, softly, very softly, and secretly.

"But what does all this mean," said Wanda, resting her head in both hands with her gaze lost in the distance, "a golden fancy which never can become true." An uncanny brooding melancholy seemed shed over her entire being; I have never seen her like that.

"Why unachievable?" I began.

"Because slavery doesn't exist any longer."

"Then we will go to a country where it still exists, to the Orient, to Turkey," I said eagerly.

"You would--Severin--in all seriousness," Wanda replied. Her eyes burned.

"Yes, in all seriousness, I want to be your slave," I continued. "I want your power over me to be sanctified by law; I want my life to be in your hands, I want nothing that could protect or save me from you. Oh, what a voluptuous joy when once I feel myself entirely dependent upon your absolute will, your whim, at your beck and call. And then what happiness, when at some time you deign to be gracious, and the slave may kiss the lips which mean life and death to him." I knelt down, and leaned my burning forehead against her knee.

"You are talking as in a fever," said Wanda agitatedly, "and you really love me so endlessly." She held me to her breast, and covered me with kisses.

"You really want it?"

"I swear to you now by God and my honor, that I shall be your slave, wherever and whenever you wish it, as soon as you command," I exclaimed, hardly master of myself.

"And if I take you at your word?" said Wanda.

"Please do!"

"All this appeals to me," she said then. "It is different from anything else--to know that a man who worships me, and whom I love with all my heart, is so wholly mine, dependent on my will and caprice, my possession and slave, while I--"

She looked strangely at me.

"If I should become frightfully frivolous you are to blame," she continued. "It almost seems as if you were afraid of me already, but you have sworn."

"And I shall keep my oath."

"I shall see to that," she replied. "I am beginning to enjoy it, and, heaven help me, we won't stick to fancies now. You shall become my slave, and I--I shall try to be *Venus in Furs*."

* * * * *

I thought that at last I knew this woman, understood her, and now I see I have to begin at the very beginning again. Only a little while ago her reaction to my dreams was violently hostile, and now she tries to carry them into execution with the soberest seriousness.

She has drawn up a contract according to which I give my word of honor and agree under oath to be her slave, as long as she wishes.

With her arm around my neck she reads this, unprecedented, incredible document to me. The end of each sentence she punctuates with a kiss.

"But all the obligations in the contract are on my side," I said, teasing her.

"Of course," she replied with great seriousness, "you cease to be my lover, and consequently I am released from all duties and obligations towards you. You will have to look upon my favors as pure benevolence. You no longer have any rights, and no longer can lay claim to any. There can be no limit to my power over you. Remember, that you won't be much better than a dog, or some inanimate object. You will be mine, my plaything, which I can break to pieces, whenever I want an hour's amusement. You are nothing, I am everything. Do you understand?" She laughed and kissed me again, and yet a sort of cold shiver ran through me.

"Won't you allow me a few conditions--" I began.

"Conditions?" She contracted her forehead. "Ah! You are afraid already, or perhaps you regret, but it is too late now. You have sworn, I have your word of honor. But let me hear them."

"First of all I should like to have it included in our contract, that you will never completely leave me, and then that you will never give me over to the mercies of any of your admirers--"

"But Severin," exclaimed Wanda with her voice full of emotion and with tears in her eyes, "how can you imagine that I--and you, a man who loves me so absolutely, who puts himself so entirely in my power--" She halted.

"No, no!" I said, covering her hands with kisses. "I don't fear anything from you that might dishonor me. Forgive me the ugly thought."

Wanda smiled happily, leaned her cheek against mine, and seemed to reflect.

"You have forgotten something," she whispered coquettishly, "the most important thing!"

"A condition?"

"Yes, that I must always wear my furs," exclaimed Wanda. "But I promise you I'll do that anyhow because they give me a despotic feeling. And I shall be very cruel to you, do you understand?"

"Shall I sign the contract?" I asked.

"Not yet," said Wanda. "I shall first add your conditions, and the actual signing won't occur until the proper time and place."

"In Constantinople?"

"No. I have thought things over. What special value would there be in owning a slave where everyone owns slaves. What I want is to *have a slave, I alone,* here in our civilized sober, Philistine world, and a slave who submits helplessly to my power solely on account of my beauty and personality, not because of law, of property rights, or compulsions. This attracts me. But at any rate we will go to a country where we are not known and where you can appear before the world as my servant without embarrassment. Perhaps to Italy, to Rome or Naples."

* * * * *

We were sitting on Wanda's ottoman. She wore her ermine jacket, her hair was loose and fell like a lion's mane down her back. She clung to my lips, drawing my soul from my body. My head whirled, my blood began to seethe, my heart beat violently against hers.

"I want to be absolutely in your power, Wanda," I exclaimed suddenly, seized by that frenzy of passion when I can scarcely think clearly or decide freely. "I want to put myself absolutely at your mercy for good or evil without any condition, without any limit to your power."

While saying this I had slipped from the ottoman, and lay at her feet looking up at her with drunken eyes.

"How beautiful you now are," she exclaimed, "your eyes half-broken in ecstacy fill me with joy, carry me away. How wonderful your look would be if you were being beaten to death, in the extreme agony. You have the eye of a martyr."

* * * * *

Sometimes, nevertheless, I have an uneasy feeling about placing myself so absolutely, so unconditionally into a woman's hands. Suppose she did abuse my passion, her power?

Well, then I would experience what has occupied my imagination since my childhood, what has always given me the feeling of seductive terror. A foolish apprehension! It will be a wanton game she will play with me, nothing more. She loves me, and she is good, a noble personality, incapable of a breach of faith. But it lies in her hands --*if she wants to she can*. What a temptation in this doubt, this fear!

Now I understand Manon l'Escault and the poor chevalier, who, even in the pillory, while she was another man's mistress, still adored her.

Love knows no virtue, no profit; it loves and forgives and suffers everything, because it must. It is not our judgment that leads us; it is neither the advantages nor the faults which we discover, that make us abandon ourselves, or that repel us.

It is a sweet, soft, enigmatic power that drives us on. We cease to think, to feel, to will; we let ourselves be carried away by it, and ask not whither?

* * * * *

A Russian prince made his first appearance today on the promenade. He aroused general interest on account of his athletic figure, magnificent face, and splendid bearing. The women particularly gaped at him as though he were a wild animal, but he went his way gloomily without paying attention to any one. He was accompanied by two servants, one a negro, completely dressed in red satin, and the other a Circassian in his full gleaming uniform. Suddenly he saw Wanda, and fixed his cold piercing look upon her; he even turned his head after her, and when she had passed, he stood still and followed her with his eyes.

And she--she veritably devoured him with her radiant green eyes--and did everything possible to meet him again.

The cunning coquetry with which she walked, moved, and looked at him, almost stifled me. On the way home I remarked about it. She knit her brows.

"What do you want," she said, "the prince is a man whom I might like, who even dazzles me, and I am free. I can do what I please--"

"Don't you love me any longer--" I stammered, frightened.

"I love only you," she replied, "but I shall have the prince pay court to me."

"Wanda!"

"Aren't you my slave?" she said calmly. "Am I not Venus, the cruel northern Venus in Furs?"

I was silent. I felt literally crushed by her words; her cold look entered my heart like a dagger.

"You will find out immediately the prince's name, residence, and circumstances," she continued. "Do you understand?"

"But--"

"No argument, obey!" exclaimed Wanda, more sternly than I would have thought possible for her, "and don't dare to enter my sight until you can answer my questions."

It was not till afternoon that I could obtain the desired information for Wanda. She let me stand before her like a servant, while she leaned back in her arm-chair and listened to me, smiling. Then she nodded; she seemed to be satisfied.

"Bring me my footstool," she commanded shortly.

I obeyed, and after having put it before her and having put her feet on it, I remained kneeling.

"How will this end?" I asked sadly after a short pause.

She broke into playful laughter. "Why things haven't even begun yet."

"You are more heartless than I imagined," I replied, hurt.

"Severin," Wanda began earnestly. "I haven't done anything yet, not the slightest thing, and you are already calling me heartless. What will happen when I begin to carry your dreams to their realization, when I shall lead a gay, free life and have a circle of admirers about me, when I shall actually fulfil your ideal, tread you underfoot and apply the lash?"

"You take my dreams too seriously."

"Too seriously? I can't stop at make-believe, when once I begin," she replied. "You know I hate all play-acting and comedy. You have wished it. Was it my idea or yours? Did I persuade you or did you inflame my imagination? I am taking things seriously now."

"Wanda," I replied, caressingly, "listen quietly to me. We love each other infinitely, we are very happy, will you sacrifice our entire future to a whim?"

"It is no longer a whim," she exclaimed.

"What is it?" I asked frightened.

"Something that was probably latent in me," she said quietly and thoughtfully. "Perhaps it would never have come to light, if you had not called it to life, and made it grow. Now that it has become a powerful impulse, fills my whole being, now that I enjoy it, now that I cannot and do not want to do otherwise, now you want to back out-- you--are you a man?"

"Dear, sweet Wanda!" I began to caress her, kiss her.

"Don't--you are not a man--"

"And you," I flared up.

"I am stubborn," she said, "you know that. I haven't a strong imagination, and like you I am weak in execution. But when I make up my mind to do something, I carry it through, and the more certainly, the more opposition I meet. Leave me alone!"

She pushed me away, and got up.

"Wanda!" I likewise rose, and stood facing her.

"Now you know what I am," she continued. "Once more I warn you. You still have the choice. I am not compelling you to be my slave."

"Wanda," I replied with emotion and tears filling my eyes, "don't you know how I love you?"

Her lips quivered contemptuously.

"You are mistaken, you make yourself out worse than you are; you are
good and noble by nature--"

"What do you know about my nature," she interrupted vehemently, "you
will get to know me as I am."

"Wanda!"

"Decide, will you submit, unconditionally?"

"And if I say no."

"Then--"

She stepped close up to me, cold and contemptuous. As she stood before
me now, the arms folded across her breast, with an evil smile about her lips,
she was in fact the despotic woman of my dreams. Her expression seemed
hard, and nothing lay in her eyes that promised kindness or mercy.

"Well--" she said at last.

"You are angry," I cried, "you will punish me."

"Oh no!" she replied, "I shall let you go. You are free. I am not holding
you."

"Wanda--I, who love you so--"

"Yes, you, my dear sir, you who adore me," she exclaimed contemptuously,
"but who are a coward, a liar, and a breaker of promises. Leave me
instantly--"

"Wanda I--"

"Wretch!"

My blood rose in my heart. I threw myself down at her feet and began to cry.

"Tears, too!" She began to laugh. Oh, this laughter was frightful. "Leave me--I don't want to see you again."

"Oh my God!" I cried, beside myself. "I will do whatever you command, be your slave, a mere object with which you can do what you will--only don't send me away--I can't bear it--I cannot live without you." I embraced her knees, and covered her hand with kisses.

"Yes, you must be a slave, and feel the lash, for you are not a man," she said calmly. She said this to me with perfect composure, not angrily, not even excitedly, and it was what hurt most. "Now I know you, your dog-like nature, that adores where it is kicked, and the more, the more it is maltreated. Now I know you, and now you shall come to know me."

She walked up and down with long strides, while I remained crushed on my knees; my head was hanging supine, tears flowed from my eyes.

"Come here," Wanda commanded harshly, sitting down on the ottoman. I obeyed her command, and sat down beside her. She looked at me sombrely, and then a light suddenly seemed to illuminate the interior of her eye. Smiling, she drew me toward her breast, and began to kiss the tears out of my eyes.

* * * * *

The odd part of my situation is that I am like the bear in Lily's park. I can escape and don't want to; I am ready to endure everything as soon as she threatens to set me free.

* * * * *

If only she would use the whip again. There is something uncanny in the kindness with which she treats me. I seem like a little captive mouse with which a beautiful cat prettily plays. She is ready at any moment to tear it to pieces, and my heart of a mouse threatens to burst.

What are her intentions? What does she purpose to do with me?

* * * * *

It seems she has completely forgotten the contract, my slavehood. Or was it actually only stubbornness? And she gave up her whole plan as soon as I no longer opposed her and submitted to her imperial whim?

How kind she is to me, how tender, how loving! We are spending marvellously happy days.

To-day she had me read to her the scene between Faust and Mephistopheles, in which the latter appears as a wandering scholar. Her glance hung on me with strange pleasure.

"I don't understand," she said when I had finished, "how a man who can read such great and beautiful thoughts with such expression, and interpret them so clearly, concisely, and intelligently, can at the same time be such a visionary and supersensual ninny as you are."

"Were you pleased," said I, and kissed her forehead.

She gently stroked my brow. "I love you, Severin," she whispered. "I don't believe I could ever love any one more than you. Let us be sensible, what do you say?"

Instead of replying I folded her in my arms; a deep inward, yet vaguely sad happiness filled my breast, my eyes grew moist, and a tear fell upon her hand.

"How can you cry!" she exclaimed, "you are a child!"

* * * * *

On a pleasure drive we met the Russian prince in his carriage. He seemed to be unpleasantly surprised to see me by Wanda's side, and looked as if he wanted to pierce her through and through with his electric gray eyes. She, however, did not seem to notice him. I felt at that moment like kneeling down before her and kissing her feet. She let her glance glide over him indifferently as though he were an inanimate object, a tree, for instance, and turned to me with her gracious smile.

* * * * *

When I said good-night to her to-day she seemed suddenly unaccountably distracted and moody. What was occupying her?

"I am sorry you are going," she said when I was already standing on the threshold.

"It is entirely in your hands to shorten the hard period of my trial, to cease tormenting me--" I pleaded.

"Do you imagine that this compulsion isn't a torment for me, too," Wanda interjected.

"Then end it," I exclaimed, embracing her, "be my wife."

"*Never, Severin*," she said gently, but with great firmness.

"What do you mean?"

I was frightened in my innermost soul.

"*You are not the man for me.*"

I looked at her, and slowly withdrew my arm which was still about her waist; then I left the room, and she--she did not call me back.

* * * * *

A sleepless night; I made countless decisions, only to toss them aside again. In the morning I wrote her a letter in which I declared our relationship dissolved. My hand trembled when I put on the seal, and I burned my fingers.

As I went upstairs to hand it to the maid, my knees threatened to give way.

The door opened, and Wanda thrust forth her head full of curling- papers.

"I haven't had my hair dressed yet," she said, smiling. "What have you there?"

"A letter--"

"For me?"

I nodded.

"Ah, you want to break with me," she exclaimed, mockingly.

"Didn't you tell me yesterday that I wasn't the man for you?"

"I repeat it now!"

"Very well, then." My whole body was trembling, my voice failed me, and I handed her the letter.

"Keep it," she said, measuring me coldly. "You forget that is no longer a question as to whether you satisfy me as a man; as a *slave* you will doubtless do well enough."

"Madame!" I exclaimed, aghast.

"That is what you will call me in the future," replied Wanda, throwing back her head with a movement of unutterable contempt. "Put your affairs in order within the next twenty-four hours. The day after to-morrow I shall start for Italy, and you will accompany me as my servant."

"Wanda--"

"I forbid any sort of familiarity," she said, cutting my words short, "likewise you are not to come in unless I call or ring for you, and you are not to speak to me until you are spoken to. From now on your name is no longer Severin, but *Gregor*."

I trembled with rage, and yet, unfortunately, I cannot deny it, I also felt a strange pleasure and stimulation.

"But, madame, you know my circumstances," I began in my confusion. "I am dependent on my father, and I doubt whether he will give me the large sum of money needed for this journey--"

"That means you have no money, Gregor," said Wanda, delightedly, "so much the better, you are then entirely dependent on me, and in fact my slave."

"You don't consider," I tried to object, "that as man of honor it is impossible for me--"

"I have indeed considered it," she replied almost with a tone of command. "As a man of honor you must keep your oath and redeem your promise to follow me as slave whithersoever I demand and to obey whatever I command. Now leave me, Gregor!"

I turned toward the door.

"Not yet--you may first kiss my hand." She held it out to me with a certain proud indifference, and I the dilettante, the donkey, the miserable slave pressed it with intense tenderness against my lips which were dry and hot with excitement.

There was another gracious nod of the head.

Then I was dismissed.

* * * * *

Though it was late in the evening my light was still lit, and a fire was
burning in the large green stove. There were still many things among my
letters and documents to be put in order. Autumn, as is usually the case
with us, had fallen with all its power.

Suddenly she knocked at my window with the handle of her whip.

I opened and saw her standing outside in her ermine-lined jacket and in a
high round Cossack cap of ermine of the kind which the great Catherine
favored.

"Are you ready, Gregor?" she asked darkly.

"Not yet, mistress," I replied.

"I like that word," she said then, "you are always to call me mistress, do
you understand? We leave here to-morrow morning at nine o'clock. As far
as the district capital you will be my companion and friend, but from the
moment that we enter the railway-coach you are my slave, my servant.
Now close the window, and open the door."

After I had done as she had demanded, and after she had entered, she asked,
contracting her brows ironically, "well, how do you like me."

"Wanda, you--"

"Who gave you permission?" She gave me a blow with the whip.

"You are very beautiful, mistress."

Wanda smiled and sat down in the arm-chair. "Kneel down--here beside my chair."

I obeyed.

"Kiss my hand."

I seized her small cold hand and kissed it.

"And the mouth--"

In a surge of passion I threw my arms around the beautiful cruel woman, and covered her face, arms, and breast with glowing kisses. She returned them with equal fervor--the eyelids closed as in a dream. It was after midnight when she left.

* * * * *

At nine o'clock sharp in the morning everything was ready for departure, as she had ordered. We left the little Carpathian health- resort in a comfortable light carriage. The most interesting drama of my life had reached a point of development whose denouement it was then impossible to foretell.

So far everything went well. I sat beside Wanda, and she chatted very graciously and intelligently with me, as with a good friend, concerning Italy, Pisemski's new novel, and Wagner's music. She wore a sort of Amazonesque travelling-dress of black cloth with a short jacket of the same material, set with dark fur. It fitted closely and showed her figure to best advantage. Over it she wore dark furs. Her hair wound into an antique knot, lay beneath a small dark fur-hat from which a black veil hung. Wanda was in very good humor; she fed me candies, played with my hair, loosened my neck cloth and made a pretty cockade of it; she covered my knees with her furs and stealthily pressed the fingers of my hand. When our Jewish driver persistently went on nodding to himself, she even gave me a kiss, and her cold lips had the fresh frosty fragrance of a young autumnal rose, which blossoms alone amid bare stalks and yellow leaves and upon whose calyx the first frost has hung tiny diamonds of ice.

* * * * *

We are at the district capital. We get out at the railway station. Wanda throws off her furs and places them over my arm, and goes to secure the tickets.

When she returns she has completely changed.

"Here is your ticket, Gregor," she says in a tone which supercilious ladies use to their servants.

"A third-class ticket," I reply with comic horror.

"Of course," she continues, "but now be careful. You won't get on until I am settled in my compartment and don't need you any longer. At each station you will hurry to my car and ask for my orders. Don't forget. And now give me my furs."

After I had helped her into them, humbly like a slave, she went to find an empty first-class coupe. I followed. Supporting herself on my shoulder, she got on and I wrapped her feet in bear-skins and placed them on the warming bottle.

Then she nodded to me, and dismissed me. I slowly ascended a third- class carriage, which was filled with abominable tobacco-smoke that seemed like the fogs of Acheron at the entrance to Hades. I now had the leisure to muse about the riddle of human existence, and about its greatest riddle of all--*woman.*

* * * * *

Whenever the train stops, I jump off, run to her carriage, and with drawn cap await her orders. She wants coffee and then a glass of water, at another time a bowl of warm water to wash her hands, and thus it goes on. She lets several men who have entered her compartment pay court to her. I am dying of jealousy and have to leap about like an antelope so as to secure what she wants quickly and not miss the train.

In this way the night passes. I haven't had time to eat a mouthful and I can't sleep, I have to breathe the same oniony air with Polish peasants, Jewish peddlers, and common soldiers.

When I mount the steps of her coupe, she is lying stretched out on cushions in her comfortable furs, covered up with the skins of animals. She is like an oriental despot, and the men sit like Indian deities, straight upright against the walls and scarcely dare to breathe.

* * * * *

She stops over in Vienna for a day to go shopping, and particularly to buy series of luxurious gowns. She continues to treat me as her servant. I follow her at the respectful distance of ten paces. She hands me her packages without so much as even deigning a kind look, and laden down like a donkey I pant along behind.

Before leaving she takes all my clothes and gives them to the hotel waiters. I am ordered to put on her livery. It is a Cracovian costume in her colors, light-blue with red facings, and red quadrangular cap, ornamented with peacock-feathers. The costume is rather becoming to me.

The silver buttons bear her coat of arms. I have the feeling of having been sold or of having bonded myself to the devil. My fair demon leads me from Vienna to Florence. Instead of linen-garbed Mazovians and greasy-haired Jews, my companions now are curly- haired Contadini, a magnificent sergeant of the first Italian Grenadiers, and a poor German painter. The tobacco smoke no longer smells of onions, but of salami and cheese.

Night has fallen again. I lie on my wooden bed as on a rack; my arms and legs seem broken. But there nevertheless is an element of poetry in the affair. The stars sparkle round about, the Italian sergeant has a face like Apollo Belvedere, and the German painter sings a lovely German song.

"Now that all the shadows gather And endless stars grow light, Deep yearning on me falls And softly fills the night."

"Through the sea of dreams Sailing without cease, Sailing goes my soul In thine to find release."

And I am thinking of the beautiful woman who is sleeping in regal comfort among her soft furs.

* * * * *

Florence! Crowds, cries, importunate porters and cab-drivers. Wanda chooses a carriage, and dismisses the porters.

"What have I a servant for," she says, "Gregor--here is the ticket-- get the luggage."

She wraps herself in her furs and sits quietly in the carriage while I drag the heavy trunks hither, one after another. I break down for a moment under the last one; a good-natured *carabiniere* with an intelligent face comes to my assistance. She laughs.

"It must be heavy," said she, "all my furs are in it."

I get up on the driver's seat, wiping drops of perspiration from my brow. She gives the name of the hotel, and the driver urges on his horse. In a few minutes we halt at the brilliantly illuminated entrance.

"Have you any rooms?" she asks the portier.

"Yes, madame."

"Two for me, one for my servant, all with stoves."

"Two first-class rooms for you, madame, both with stoves," replied the waiter who had hastily come up, "and one without heat for your servant."

She looked at them, and then abruptly said: "they are satisfactory, have fires built at once; my servant can sleep in the unheated room."

I merely looked at her.

"Bring up the trunks, Gregor," she commands, paying no attention to my looks. "In the meantime I'll be dressing, and then will go down to the dining-room, and you can eat something for supper."

As she goes into the adjoining room, I drag the trunks upstairs and help the waiter build a fire in her bed-room. He tries to question me in bad French about my employer. With a brief glance I see the blazing fire, the fragrant white poster-bed, and the rugs which cover the floor. Tired and hungry I then descend the stairs, and ask for something to eat. A good-natured waiter, who used to be in the Austrian army and takes all sorts of pains to entertain me in German, shows me the dining-room and waits on me. I have just had the first fresh drink in thirty-six hours and the first bite of warm food on my fork, when she enters.

I rise.

"What do you mean by taking me into a dining-room in which my servant is eating," she snaps at the waiter, flaring with anger. She turns around and leaves.

Meanwhile I thank heaven that I am permitted to go on eating. Later I climb the four flights upstairs to my room. My small trunk is already there, and a miserable little oil-lamp is burning. It is a narrow room without fire-place, without a window, but with a small air-hole. If it weren't so beastly cold, it would remind me of one of the Venetian *piombi*. [Footnote: These were notorious prisons under the leaden roof of the Palace of the

Doges.] Involuntarily I have to laugh out aloud, so that it re-echoes, and I am startled by my own laughter.

Suddenly the door is pulled open and the waiter with a theatrical Italian gesture calls "You are to come down to madame, at once." I pick up my cap, stumble down the first few steps, but finally arrive in front of her door on the first floor and knock.

"Come in!"

I enter, shut the door, and stand attention.

Wanda has made herself comfortable. She is sitting in a neglige of white muslin and laces on a small red divan with her feet on a footstool that matches. She has thrown her fur-cloak about her. It is the identical cloak in which she appeared to me for the first time, as goddess of love.

The yellow lights of the candelabra which stand on projections, their reflections in the large mirrors, and the red flames from the open fireplace play beautifully on the green velvet, the dark-brown sable of the cloak, the smooth white skin, and the red, flaming hair of the beautiful woman. Her clear, but cold face is turned toward me, and her cold green eyes rest upon me.

"I am satisfied with you, Gregor," she began.

I bowed.

"Come closer."

I obeyed.

"Still closer," she looked down, and stroked the sable with her hand. "Venus in Furs receives her slave. I can see that you are more than an ordinary dreamer, you don't remain far in arrears of your dreams; you are the sort of man who is ready to carry his dreams into effect, no matter how mad they are. I confess, I like this; it impresses me. There is strength in this, and strength is the only thing one respects. I actually believe that under unusual circumstances, in a period of great deeds, what seems to be your weakness would reveal itself as extraordinary power. Under the early emperors you would have been a martyr, at the time of the Reformation an anabaptist, during the French Revolution one of those inspired Girondists who mounted the guillotine with the marseillaise on their lips. But you are my slave, my--"

She suddenly leaped up; the furs slipped down, and she threw her arms with soft pressure about my neck.

"My beloved slave, Severin, oh, how I love you, how I adore you, how handsome you are in your Cracovian costume! You will be cold to-night up in your wretched room without a fire. Shall I give you one of my furs, dear heart, the large one there--"

She quickly picked it up, throwing it over my shoulders, and before I knew what had happened I was completely wrapped up in it.

"How wonderfully becoming furs are to your face, they bring out your noble lines. As soon as you cease being my slave, you must wear a velvet coat with sable, do you understand? Otherwise I shall never put on my fur-jacket again."

And again she began to caress me and kiss me; finally she drew me down on the little divan.

"You seem to be pleased with yourself in furs," she said. "Quick, quick, give them to me, or I will lose all sense of dignity."

I placed the furs about her, and Wanda slipped her right arm into the sleeve.

"This is the pose in Titian's picture. But now enough of joking. Don't always look so solemn, it makes me feel sad. As far as the world is concerned you are still merely my servant; you are not yet my slave, for you have not yet signed the contract. You are still free, and can leave me any moment. You have played your part magnificently. I have been delighted, but aren't you tired of it already, and don't you think I am abominable? Well, say something--I command it."

"Must I confess to you, Wanda?" I began.

"Yes, you must."

"Even it you take advantage of it," I continued, "I shall love you the more deeply, adore you the more fanatically, the worse you treat me. What you have just done inflames my blood and intoxicates all my senses." I held her close to me and clung for several moments to her moist lips.

"Oh, you beautiful woman," I then exclaimed, looking at her. In my enthusiasm I tore the sable from her shoulders and pressed my mouth against her neck.

"You love me even when I am cruel," said Wanda, "now go!--you bore me--don't you hear?"

She boxed my ears so that I saw stars and bells rang in my ears.

"Help me into my furs, slave."

I helped her, as well as I could.

"How awkward," she exclaimed, and was scarcely in it before she struck me in the face again. I felt myself growing pale.

"Did I hurt you?" she asked, softly touching me with her hand.

"No, no," I exclaimed.

"At any rate you have no reason to complain, you want it thus; now kiss me again."

I threw my arms about her, and her lips clung closely to mine. As she lay against my breast in her large heavy furs, I had a curiously oppressive sensation. It was as if a wild beast, a she-bear, were embracing me. It seemed as if I were about to feel her claws in my flesh. But this time the she-bear let me off easily.

With my heart filled with smiling hopes, I went up to my miserable servant's room, and threw myself down on my hard couch.

"Life is really amazingly droll," I thought. "A short time ago the most beautiful woman, Venus herself, rested against your breast, and now you have an opportunity for studying the Chinese hell. Unlike us, they don't hurl the damned into flames, but they have devils chasing them out into fields of ice.

"Very likely the founders of their religion also slept in unheated rooms."

* * * * *

During the night I startled out of my sleep with a scream. I had been dreaming of an icefield in which I had lost my way; I had been looking in vain for a way out. Suddenly an eskimo drove up in a sleigh harnessed with reindeer; he had the face of the waiter who had shown me to the unheated room.

"What are you looking for here, my dear sir?" he exclaimed. "This is the North Pole."

A moment later he had disappeared, and Wanda flew over the smooth ice on tiny skates. Her white satin skirt fluttered and crackled; the ermine of her jacket and cap, but especially her face, gleamed whiter than the snow. She shot toward me, inclosed me in her arms, and began to kiss me. Suddenly I felt my blood running warm down my side.

"What are you doing?" I asked horror-stricken.

She laughed, and as I looked at her now, it was no longer Wanda, but a huge, white she-bear, who was digging her paws into my body.

I cried out in despair, and still heard her diabolical laughter when I awoke, and looked about the room in surprise.

Early in the morning I stood at Wanda's door, and the waiter brought the coffee. I took it from him, and served it to my beautiful mistress. She had already dressed, and looked magnificent, all fresh and roseate. She smiled graciously at me and called me back, when I was about to withdraw respectfully.

"Come, Gregor, have your breakfast quickly too," she said, "then we will go house-hunting. I don't want to stay in the hotel any longer than I have to. It is very embarassing here. If I chat with you for more than a minute, people will immediately say: 'The fair Russian is having an affair with her servant, you see, the race of Catherines isn't extinct yet.'"

Half an hour later we went out; Wanda was in her cloth-gown with the Russian cap, and I in my Cracovian costume. We created quite a stir. I walked about ten paces behind, looking very solemn, but expected momentarily to have to break out into loud laughter. There was scarcely a street in which one or the other of the attractive houses did not bear the sign *camere ammobiliate*. Wanda always sent me upstairs, and only when the apartment seemed to answer her requirements did she herself ascend. By noon I was as tired as a stag- hound after the hunt.

We entered a new house and left it again without having found a suitable habitation. Wanda was already somewhat out of humor. Suddenly she said to me: "Severin, the seriousness with which you play your part is charming, and the restrictions, which we have placed upon each other are really annoying me. I can't stand it any longer, I do love you, I must kiss you. Let's go into one of the houses."

"But, my lady--" I interposed.

"Gregor?" She entered the next open corridor and ascended a few steps of the dark stair-way; then she threw her arms about me with passionate tenderness and kissed me.

"Oh, Severin, you were very wise. You are much more dangerous as slave than I would have imagined; you are positively irrestible, and I am afraid I shall have to fall in love with you again."

"Don't you love me any longer then," I asked seized by a sudden fright.

She solemnly shook her head, but kissed me again with her swelling, adorable lips.

We returned to the hotel. Wanda had luncheon, and ordered me also quickly to get something to eat.

Of course, I wasn't served as quickly as she, and so it happened that just as I was carrying the second bite of my steak to my mouth, the waiter entered and called out with his theatrical gesture: "Madame wants you, at once."

I took a rapid and painful leave of my food, and, tired and hungry, hurried toward Wanda, who was already on the street.

"I wouldn't have imagined you could be so cruel," I said reproachfully. "With all these, fatiguing duties you don't even leave me time to eat in peace."

Wanda laughed gaily. "I thought you had finished," she said, "but never mind. Man was born to suffer, and you in particular. The martyrs didn't have any beefsteaks either."

I followed her resentfully, gnawing at my hunger.

"I have given up the idea of finding a place in the city," Wanda continued. "It will be difficult to find an entire floor which is shut off and where you can do as you please. In such a strange, mad relationship as ours there must be no jarring note. I shall rent an entire villa--and you will be surprised. You have my permission now to satisfy your hunger, and look about a bit in Florence. I won't be home till evening. If I need you then, I will have you called."

I looked at the Duomo, the Palazzo Vecchio, the Logia di Lanzi, and then I stood for a long time on the banks of the Arno. Again and again I let my eyes rest on the magnificent ancient Florence, whose round cupolas and towers were drawn in soft lines against the blue, cloudless sky. I watched its splendid bridges beneath whose wide arches the lively waves of the beautiful, yellow river ran, and the green hills which surrounded the city, bearing slender cypresses and extensive buildings, palaces and monasteries.

It is a different world, this one in which we are--a gay, sensuous, smiling world. The landscape too has nothing of the seriousness and somberness of ours. It is a long ways off to the last white villas scattered among the pale green of the mountains, and yet there isn't a spot that isn't bright with sunlight. The people are less serious than we; perhaps, they think less, but they all look as though they were happy.

It is also maintained that death is easier in the South.

I have a vague feeling now that such a thing as beauty without thorn and love of the senses without torment does exist.

Wanda has discovered a delightful little villa and rented it for the winter. It is situated on a charming hill on the left bank of the Arno, opposite the Cascine. It is surrounded by an attractive garden with lovely paths, grass plots, and magnificent meadow of camelias. It is only two stories high, quadrangular in the Italian fashion. An open gallery runs along one side, a sort of loggia with plaster-casts of antique statues; stone steps lead from it down into the garden. From the gallery you enter a bath with a magnificent marble basin, from which winding stairs lead to my mistress' bed-chamber.

Wanda occupies the second story by herself.

A room on the ground floor has been assigned to me; it is very attractive, and even has a fireplace.

I have roamed through the garden. On a round hillock I discovered a little temple, but I found its door locked. However, there is a chink in the door and when I glue my eye to it, I see the goddess of love on a white pedestal.

A slight shudder passes over me. It seems to me as if she were smiling at me saying: "Are you there? I have been expecting you."

* * * * *

It is evening. An attractive maid brings me orders to appear before my mistress. I ascend the wide marble stairs, pass through the anteroom, a large salon furnished with extravagant magnificence, and knock at the door of the bedroom. I knock very softly for the luxury displayed everywhere intimidates me. Consequently no one hears me, and I stand for some time in front of the door. I have a feeling as if I were standing before the bed-room of the great Catherine, and it seems as if at any moment she might come out in her green sleeping furs, with the red ribbon and decoration on her bare breast, and with her little white powdered curls.

I knocked again. Wanda impatiently pulls the door open.

"Why so late?" she asks.

"I was standing in front of the door, but you didn't hear me knock," I reply timidly. She closes the door, and clinging to me, she leads me to the red damask ottoman on which she had been resting. The entire arrangement of the room is in red damask--wall-paper, curtains, portieres, hangings of the bed. A magnificent painting of Samson and Delilah forms the ceiling.

Wanda receives me in an intoxicating dishabille. Her white satin dress flows gracefully and picturesquely down her slender body, leaving her arms and breast bare, and carelessly they nestle amid the dark hair of the great fur of sable, lined with green velvet. Her red hair falls down her back as far as the hips, only half held by strings of black pearls.

"Venus in Furs," I whisper, while she draws me to her breast and threatens to stifle me with her kisses. Then I no longer speak and neither do I think; everything is drowned out in an ocean of unimagined bliss.

"Do you still love me?" she asks, her eye softening in passionate tenderness.

"You ask!" I exclaimed.

"You still remember your oath," she continued with an alluring smile, "now that everything is prepared, everything in readiness, I ask you once more, is it still your serious wish to become my slave?"

"Am I not ready?" I asked in surprise.

"You have not yet signed the papers."

"Papers--what papers?"

"Oh, I see, you want to give it up," she said, "well then, we will let it go."

"But Wanda," I said, "you know that nothing gives me greater happiness than to serve you, to be your slave. I would give everything for the sake of feeling myself wholly in your power, even unto death--"

"How beautiful you are," she whispered, "when you speak so enthusiastically, so passionately. I am more in love with you than ever and you want me to be dominant, stern, and cruel. I am afraid, it will be impossible for me to be so."

"I am not afraid," I replied smiling, "where are the papers?'"

"So that you may know what it means to be absolutely in my power, I have drafted a second agreement in which you declare that you have decided to kill yourself. In that way I can even kill you, if I so desire."

"Give them to me."

While I was unfolding the documents and reading them, Wanda got pen and ink. She then sat down beside me with her arm about my neck, and looked over my shoulder at the paper.

The first one read:

AGREEMENT BETWEEN MME. VON DUNAJEW AND SEVERIN VON KUSIEMSKI

"Severin von Kusiemski ceases with the present day being the affianced of Mme. Wanda von Dunajew, and renounces all the rights appertaining thereunto; he on the contrary binds himself on his word of honor as a man and nobleman, that hereafter he will be her *slave* until such time that she herself sets him at liberty again.

"As the slave of Mme. von Dunajew he is to bear the name Gregor, and he is unconditionally to comply with every one of her wishes, and to obey every one of her commands; he is always to be submissive to his mistress, and is to consider her every sign of favor as an extraordinary mercy.

"Mme. von Dunajew is entitled not only to punish her slave as she deems best, even for the slightest inadvertence or fault, but also is herewith given the right to torture him as the mood may seize her or merely for the sake of whiling away the time. Should she so desire, she may kill him whenever she wishes; in short, he is her unrestricted property.

"Should Mme. von Dunajew ever set her slave at liberty, Severin von Kusiemski agrees to forget everything that he has experienced or suffered as her slave, and promises *never under any circumstances and in no wise to think of vengeance or retaliation.*

"Mme. von Dunajew on her behalf agrees as his mistress to appear as often as possible in her furs, especially when she purposes some cruelty toward her slave."

Appended at the bottom of the agreement was the date of the present day.

The second document contained only a few words.

"Having since many years become weary of existence and its illusions, I have of my own free will put an end to my worthless life."

I was seized with a deep horror when I had finished. There was still time, I could still withdraw, but the madness of passion and the sight of the beautiful woman that lay all relaxed against my shoulder carried me away.

"This one you will have to copy, Severin," said Wanda, indicating the second document. "It has to be entirely in your own handwriting; this, of course, isn't necessary in the case of the agreement."

I quickly copied the few lines in which I designated myself a suicide, and handed them to Wanda. She read them, and put them on the table with a smile.

"Now have you the courage to sign it?" she asked with a crafty smile, inclining her head.

I took the pen.

"Let me sign first," said Wanda, "your hand is trembling, are you afraid of the happiness that is to be yours?"

She took the agreement and pen. While engaging in my internal struggle, I looked upward for a moment. It occurred to me that the painting on the ceiling, like many of those of the Italian and Dutch schools, was utterly unhistorical, but this very fact gave it a strange mood which had an almost uncanny effect on me. Delilah, an opulent woman with flaming red hair, lay extended, half-disrobed, in a dark fur-cloak, upon a red ottoman, and bent smiling over Samson who had been overthrown and bound by the Philistines. Her smile in its mocking coquetry was full of a diabolical cruelty; her eyes, half- closed, met Samson's, and his with a last look of insane passion cling to hers, for already one of his enemies is kneeling on his breast with the red-hot iron to blind him.

"Now--" said Wanda. "Why you are all lost in thought. What is the matter with you, everything will remain just as it was, even after you have signed, don't you know me yet, dear heart?"

I looked at the agreement. Her name was written there in bold letters. I peered once more into her eyes with their potent magic, then I took the pen and quickly signed the agreement.

"You are trembling," said Wanda calmly, "shall I help you?"

She gently took hold of my hand, and my name appeared at the bottom of the second paper. Wanda looked once more at the two documents, and then locked them in the desk which stood at the head of the ottoman.

"Now then, give me your passport and money."

I took out my wallet and handed it to her. She inspected it, nodded, and put it with other things while in a sweet drunkenness I kneeled before her leaning my head against her breast.

Suddenly she thrusts me away with her foot, leaps up, and pulls the bell-rope. In answer to its sound three young, slender negresses enter; they are as if carved of ebony, and are dressed from head to foot in red satin; each one has a rope in her hand.

Suddenly I realize my position, and am about to rise. Wanda stands proudly erect, her cold beautiful face with its sombre brows and contemptous eyes is turned toward me. She stands before me as mistress, commanding, gives a sign with her hand, and before I really know what has happened to me the

negresses have dragged me to the ground, and have tied me hand and foot.
As in the case of one about to be executed my arms are bound behind my
back, so that I can scarcely move.

"Give me the whip, Haydee," commands Wanda, with unearthly calm.

The negress hands it to her mistress, kneeling.

"And now take off my heavy furs," she continues, "they impede me."

The negress obeyed.

"The jacket there!" Wanda commanded.

Haydee quickly brought her the *kazabaika*, set with ermine, which lay on
the bed, and Wanda slipped into it with two inimitably graceful
movements.

"Now tie him to the pillar here!"

The negresses lifted me up, and twisting a heavy rope around my body, tied
me standing against one of the massive pillars which supported the top of
the wide Italian bed.

Then they suddenly disappeared, as if the earth had swallowed them.

Wanda swiftly approached me. Her white satin dress flowed behind her in a
long train, like silver, like moonlight; her hair flared like flames against the
white fur of her jacket. Now she stood in front of me with her left hand
firmly planted on her hips, in her right hand she held the whip. She uttered
an abrupt laugh.

"Now play has come to an end between us," she said with heartless
coldness. "Now we will begin in dead earnest. You fool, I laugh at you and
despise you; you who in your insane infatuation have given yourself as a
plaything to *me*, the frivolous and capricious woman. You are no longer the

man I love, but *my slave*, at my mercy even unto life and death.

"You shall know me!

"First of all you shall have a taste of the whip in all seriousness, without having done anything to deserve it, so that you may understand what to expect, if you are awkward, disobedient, or refractory."

With a wild grace she rolled back her fur-lined sleeve, and struck me across the back.

I winced, for the whip cut like a knife into my flesh.

"Well, how do you like that?" she exclaimed.

I was silent.

"Just wait, you will yet whine like a dog beneath my whip," she threatened, and simultaneously began to strike me again.

The blows fell quickly, in rapid succession, with terrific force upon my back, arms, and neck; I had to grit my teeth not to scream aloud. Now she struck me in the face, warm blood ran down, but she laughed, and continued her blows.

"It is only now I understand you," she exclaimed. "It really is a joy to have some one so completely in one's power, and a man at that, who loves you--you do love me?--No--Oh! I'll tear you to shreds yet, and with each blow my pleasure will grow. Now, twist like a worm, scream, whine! You will find no mercy in me!"

Finally she seemed tired.

She tossed the whip aside, stretched out on the ottoman, and rang.

The negresses entered.

"Untie him!"

As they loosened the rope, I fell to the floor like a lump of wood. The black women grinned, showing their white teeth.

"Untie the rope around his feet."

They did it, but I was unable to rise.

"Come over here, Gregor."

I approached the beautiful woman. Never did she seem more seductive to me than to-day in spite of all her cruelty and contempt.

"One step further," Wanda commanded. "Now kneel down, and kiss my foot."

She extended her foot beyond the hem of white satin, and I, the supersensual fool, pressed my lips upon it.

"Now, you won't lay eyes on me for an entire month, Gregor," she said seriously. "I want to become a stranger to you, so you will more easily adjust yourself to our new relationship. In the meantime you will work in the garden, and await my orders. Now, off with you, slave!"

* * * * *

A month has passed with monotonous regularity, heavy work, and a melancholy hunger, hunger for her, who is inflicting all these torments on me.

I am under the gardener's orders; I help him lop the trees and prune the hedges, transplant flowers, turn over the flower beds, sweep the gravel paths; I share his coarse food and his hard cot; I rise and go to bed with the chickens. Now and then I hear that our mistress is amusing herself, surrounded by admirers. Once I heard her gay laughter even down here in the garden.

I seem awfully stupid to myself. Was it the result of my present life, or was I so before? The month is drawing to a close--the day after to-morrow. What will she do with me now, or has she forgotten me, and left me to trim hedges and bind bouquets till my dying day?

A written order.

"The slave Gregor is herewith ordered to my personal service.

Wanda Dunajew."

With a beating heart I draw aside the damask curtain on the following morning, and enter the bed-room of my divinity. It is still filled with a pleasant half darkness.

"Is it you, Gregor?" she asks, while I kneel before the fire-place, building a fire. I tremble at the sound of the beloved voice. I cannot see her herself; she is invisible behind the curtains of the four-poster bed.

"Yes, my mistress," I reply.

"How late is it?"

"Past nine o'clock."

"Breakfast."

I hasten to get it, and then kneel down with the tray beside her bed.

"Here is breakfast, my mistress."

Wanda draws back the curtains, and curiously enough at the first glance when I see her among the pillows with loosened flowing hair, she seems an absolute stranger, a beautiful woman, but the beloved soft lines are gone.

This face is hard and has an expression of weariness and satiety.

Or is it simply that formerly my eye did not see this?

She fixes her green eyes upon me, more with curiosity than with menace, perhaps even somewhat pityingly, and lazily pulls the dark sleeping fur on which she lies over the bared shoulder.

At this moment she is very charming, very maddening, and I feel my blood rising to my head and heart. The tray in my hands begins to sway. She notices it and reached out for the whip which is lying on the toilet-table.

"You are awkward, slave," she says furrowing her brow.

I lower my looks to the ground, and hold the tray as steadily as possible. She eats her breakfast, yawns, and stretches her opulent limbs in the magnificent furs.

She has rung. I enter.

"Take this letter to Prince Corsini."

I hurry into the city, and hand the letter to the Prince. He is a handsome young man with glowing black eyes. Consumed with jealousy, I take his answer to her.

"What is the matter with you?" she asks with lurking spitefulness. "You are very pale."

"Nothing, mistress, I merely walked rather fast."

At luncheon the prince is at her side, and I am condemned to serve both her and him. They joke, and I am, as if non-existent, for both. For a brief moment I see black; I was just pouring some Bordeaux into his glass, and spilled it over the table-cloth and her gown.

"How awkward," Wanda exclaimed and slapped my face. The prince laughed, and she also, but I felt the blood rising to my face.

After luncheon she drove in the Cascine. She has a little carriage with a handsome, brown English horse, and holds the reins herself. I sit behind and notice how coquettishly she acts, and nods with a smile when one of the distinguished gentlemen bows to her.

As I help her out of the carriage, she leans lightly on my arm; the contact runs through me like an electric shock. She *is* a wonderful woman, and I love her more than ever.

* * * * *

For dinner at six she has invited a small group of men and women. I serve, but this time I do not spill any wine over the table-cloth.

A slap in the face is more effective than ten lectures. It makes you understand very quickly, especially when the instruction is by the way of a small woman's hand.

* * * * *

After dinner she drives to the Pergola Theater. As she descends the stairs in her black velvet dress with its large collar of ermine and with a diadem of white roses on her hair, she is literally stunning. I open the carriage-door, and help her in. In front of the theater I leap from the driver's seat, and in alighting she leaned on my arm, which trembled under the sweet burden. I open the door of her box, and then wait in the vestibule. The performance lasts four hours; she receives visits from her cavaliers, the while I grit my teeth with rage.

It is way beyond midnight when my mistress's bell sounds for the last time.

"Fire!" she orders abruptly, and when the fire-place crackles, "Tea!"

When I return with the samovar, she has already undressed, and with the aid of the negress slipped into a white negligee.

Haydee thereupon leaves.

"Hand me the sleeping-furs," says Wanda, sleepily stretching her lovely limbs. I take them from the arm-chair, and hold them while she slowly and lazily slides into the sleeves. She then throws herself down on the cushions of the ottoman.

"Take off my shoes, and put on my velvet slippers."

I kneel down and tug at the little shoe which resists my efforts. "Hurry, hurry!" Wanda exclaims, "you are hurting me! just you wait--I will teach you." She strikes me with the whip, but now the shoe is off.

"Now get out!" Still a kick--and then I can go to bed.

* * * * *

To-night I accompanied her to a soiree. In the entrance-hall she ordered me to help her out of her furs; then with a proud smile, confident of victory, she entered the brilliantly illuminated room. I again waited with gloomy and monotonous thoughts, watching hour after hour run by. From time to time the sounds of music reached me, when the door remained open for a moment. Several servants tried to start a conversation with me, but soon desisted, since I knew only a few words of Italian.

Finally I fell asleep, and dreamed that I murdered Wanda in a violent attack of jealousy. I was condemned to death, and saw myself strapped on the board; the knife fell, I felt it on my neck, but I was still alive--

Then the executioner slapped my face.

No, it wasn't the executioner; it was Wanda who stood wrathfully before me demanding her furs. I am at her side in a moment, and help her on with it.

There is a deep joy in wrapping a beautiful woman into her furs, and in seeing and feeling how her neck and magnificent limbs nestle in the precious soft furs, and to lift the flowing hair over the collar. When she throws it off a soft warmth and a faint fragrance of her body still clings to the ends of the hairs of sable. It is enough to drive one mad.

* * * * *

Finally a day came when there were neither guests, nor theater, nor other company. I breathed a sigh of relief. Wanda sat in the gallery, reading, and apparently had no orders for me. At dusk when the silvery evening mists fell she withdrew. I served her at dinner, she ate by herself, but had not a look, not a syllable for me, not even a slap in the face.

I actually desire a slap from her hand. Tears fill my eyes, and I feel that she has humiliated me so deeply, that she doesn't even find it worth while to torture or maltreat me any further.

Before she goes to bed, her bell calls me.

"You will sleep here to-night, I had horrible dreams last night, and am afraid of being alone. Take one of the cushions from the ottoman, and lie down on the bearskin at my feet."

Then Wanda put out the lights. The only illumination in the room was from a small lamp suspended from the ceiling. She herself got into bed. "Don't stir, so as not to wake me."

I did as she had commanded, but could not fall asleep for a long time. I saw the beautiful woman, beautiful as a goddess, lying on her back on the dark sleeping-furs; her arms beneath her neck, with a flood of red hair over them. I heard her magnificent breast rise in deep regular breathing, and whenever she moved ever so slightly. I woke up and listened to see whether she needed me.

But she did not require me.

No task was required of me; I meant no more to her than a night- lamp, or a revolver which one places under one's pillow.

* * * * *

Am I mad or is she? Does all this arise out of an inventive, wanton woman's brain with the intention of surpassing my supersensual fantasies, or is this woman really one of those Neronian characters who take a diabolical pleasure in treading underfoot, like a worm, human beings, who have thoughts and feelings and a will like theirs?

What have I experienced?

When I knelt with the coffee-tray beside her bed, Wanda suddenly placed her hand on my shoulder and her eyes plunged deep into mine.

"What beautiful eyes you have," she said softly, "and especially now since you suffer. Are you very unhappy?"

I bowed my head, and kept silent.

"Severin, do you still love me," she suddenly exclaimed passionately, "can you still love me?"

She drew me close with such vehemence that the coffee-tray upset, the can and cups fell to the floor, and the coffee ran over the carpet.

"Wanda--my Wanda," I cried out and held her passionately against me; I covered her mouth, face, and breast with kisses.

"It is my unhappiness that I love you more and more madly the worse you treat me, the more frequently you betray me. Oh, I shall die of pain and love and jealousy."

"But I haven't betrayed you, as yet, Severin," replied Wanda smiling.

"Not? Wanda! Don't jest so mercilessly with me," I cried. "Haven't I myself taken the letter to the Prince--"

"Of course, it was an invitation for luncheon."

"You have, since we have been in Florence--"

"I have been absolutely faithful to you" replied Wanda, "I swear it by all that is holy to me. All that I have done was merely to fulfill your dream and it was done for your sake.

"However, I shall take a lover, otherwise things will be only half accomplished, and in the end you will yet reproach me with not having treated you cruelly enough, my dear beautiful slave! But to-day you shall be Severin again, the only one I love. I haven't given away your clothes. They are here in the chest. Go and dress as you used to in the little Carpathian health-resort when our love was so intimate. Forget everything that has happened since; oh, you will forget it easily in my arms; I shall kiss away all your sorrows."

She began to treat me tenderly like a child, to kiss me and caress me. Finally she said with a gracious smile, "Go now and dress, I too will dress. Shall I put on my fur-jacket? Oh yes, I know, now run along!"

When I returned she was standing in the center of the room in her white satin dress, and the red *kazabaika* edged with ermine; her hair was white with powder and over her forehead she wore a small diamond diadem. For a moment she reminded me in an uncanny way of Catherine the Second, but she did not give me much time for reminiscences. She drew me down on the ottoman beside her and we enjoyed two blissful hours. She was no longer the stern capricious mistress, she was entirely a fine lady, a tender sweetheart. She showed me photographs and books which had just appeared, and talked about them with so much intelligence, clarity, and good taste, that I more than once carried her hand to my lips, enraptured.

She then had me recite several of Lermontov's poems, and when I was all afire with enthusiasm, she placed her small hand gently on mine. Her expression was soft, and her eyes were filled with tender pleasure.

"Are you happy?"

"Not yet."

She then leaned back on the cushions, and slowly opened her *kazabaika*.

But I quickly covered the half-bared breast again with the ermine. "You are driving me mad." I stammered.

"Come!"

I was already lying in her arms, and like a serpent she was kissing me with her tongue, when again she whispered, "Are you happy?"

"Infinitely!" I exclaimed.

She laughed aloud. It was an evil, shrill laugh which made cold shivers run down by back.

"You used to dream of being the slave, the plaything of a beautiful woman, and now you imagine you are a free human being, a man, my lover-you fool! A sign from me, and you are a slave again. Down on your knees!"

I sank down from the ottoman to her feet, but my eye still clung doubtingly on hers.

"You can't believe it," she said, looking at me with her arms folded across her breast. "I am bored, and you will just do to while away a couple of hours of time. Don't look at me that way--"

She kicked me with her foot.

"You are just what I want, a human being, a thing, an animal--"

She rang. The three negresses entered.

"Tie his hands behind his back."

I remained kneeling and unresistingly let them do this. They led me into the garden, down to the little vineyard, which forms the southern boundary. Corn had been planted between the espaliers, and here and there a few dead stalks still stood. To one side was a plough.

The negresses tied me to a post, and amused themselves sticking me with their golden hair-needles. But this did not last long, before Wanda appeared with her ermine cap on her head, and with her hands in the pockets of her jacket. She had me untied, and then my hands were fastened together on my back. She finally had a yoke put around my neck, and harnessed me to the plough.

Then her black demons drove me out into the field. One of them held the plough, the other one led me by a line, the third applied the whip, and Venus in Furs stood to one side and looked on.

* * * * *

When I was serving dinner on the following day Wanda said: "Bring another cover, I want you to dine with me to-day," and when I was about to sit down opposite her, she added, "No, over here, close by my side."

She is in the best of humors, gives me soup with her spoon, feeds me with her fork, and places her head on the table like a playful kitten and flirts with me. I have the misfortune of looking at Haydee, who serves in my place, perhaps a little longer than is necessary. It is only now that I noticed her noble, almost European cast of countenance and her magnificent statuesque bust, which is as if hewn out of black marble. The black devil observes that she pleases me, and, grinning, shows her teeth. She has hardly left the room, before Wanda leaps up in a rage.

"What, you dare to look at another woman besides me! Perhaps you like her even better than you do me, she is even more demonic!"

I am frightened; I have never seen her like this before; she is suddenly pale even to the lips and her whole body trembles. Venus in Furs is jealous of her slave. She snatches the whip from its hook and strikes me in the face; then she calls her black servants, who bind me, and carry me down into the cellar, where they throw me into a dark, dank, subterranean compartment, a veritable prison-cell.

Then the lock of the door clicks, the bolts are drawn, a key sings in the lock. I am a prisoner, buried.

I have been lying here for I don't know how long, bound like a calf about to be hauled to the slaughter, on a bundle of damp straw, without any light, without food, without drink, without sleep. It would be like her to let me starve to death, if I don't freeze to death before then. I am shaking with cold. Or is it fever? I believe I am beginning to hate this woman.

* * * * *

A red streak, like blood, floods across the floor; it is a light falling through the door which is now thrust open.

Wanda appears on the threshold, wrapped in her sables, holding a lighted torch.

"Are you still alive?" she asks.

"Are you coming to kill me?" I reply with a low, hoarse voice.

With two rapid strides Wanda reaches my side, she kneels down beside me, and places my head in her lap. "Are you ill? Your eyes glow so, do you love me? I want you to love me."

She draws forth a short dagger. I start with fright when its blade gleams in front of my eyes. I actually believe that she is about to kill me. She laughs, and cuts the ropes that bind me.

* * * * *

Every evening after dinner she now has me called. I have to read to her, and she discusses with me all sorts of interesting problems and subjects. She seems entirely transformed; it is as if she were ashamed of the savagery which she betrayed to me and of the cruelty with which she treated me. A touching gentleness transfigures her entire being, and when at the good-night she gives me her hand, a superhuman power of goodness and love lies in her eyes, of the kind which calls forth tears in us and causes us to forget all the miseries of existence and all the terrors of death.

* * * * *

I am reading *Manon l'Escault* to her. She feels the association, she doesn't say a word, but she smiles from time to time, and finally she shuts up the little book.

"Don't you want to go on reading?"

"Not to-day. We will ourselves act *Manon l'Escault* to-day. I have a rendezvous in the Cascine, and you, my dear Chevalier, will accompany me; I know, you will do it, won't you?"

"You command it."

"I do not command it, I beg it of you," she says with irresistible charm. She then rises, puts her hands on my shoulders, and looks at me.

"Your eyes!" she exclaims. "I love you, Severin, you have no idea how I love you!"

"Yes, I have!" I replied bitterly, "so much so that you have arranged for a rendezvous with some one else."

"I do this only to allure you the more," she replied vivaciously. "I must have admirers, so as not to lose you. I don't ever want to lose you, never, do you hear, for I love only you, you alone."

She clung passionately to my lips.

"Oh, if I only could, as I would, give you all of my soul in a kiss-- thus--but now come."

She slipped into a simple black velvet coat, and put a dark *bashlyk* [Footnote: A kind of Russian cap.] on her head. Then she rapidly went through the gallery, and entered the carriage.

"Gregor will drive," she called out to the coachman who withdrew in surprise.

I ascended the driver's seat, and angrily whipped up the horses.

In the Cascine where the main roadway turns into a leafy path, Wanda got out. It was night, only occasional stars shone through the gray clouds that fled across the sky. By the bank of the Arno stood a man in a dark cloak, with a brigand's hat, and looked at the yellow waves. Wanda rapidly walked through the shrubbery, and tapped him on the shoulder. I saw him turn and seize her hand, and then they disappeared behind the green wall.

An hour full of torments. Finally there was a rustling in the bushes to one side, and they returned.

The man accompanied her to the carriage. The light of the lamp fell full and glaringly upon an infinitely young, soft and dreamy face which I had never before seen, and played in his long, blond curls.

She held out her hand which he kissed with deep respect, then she signaled to me, and immediately the carriage flew along the leafy wall which follows the river like a long green screen.

* * * * *

The bell at the garden-gate rings. It is a familiar face. The man from the Cascine.

"Whom shall I announce?" I ask him in French. He timidly shakes his head.

"Do you, perhaps, understand some German?" he asks shyly.

"Yes. Your name, please."

"Oh! I haven't any yet," he replies, embarrassed--"Tell your mistress the German painter from the Cascine is here and would like-- but there she is herself."

Wanda had stepped out on the balcony, and nodded toward the stranger.

"Gregor, show the gentleman in!" she called to me.

I showed the painter the stairs.

"Thanks, I'll find her now, thanks, thanks very much." He ran up the steps. I remained standing below, and looked with deep pity on the poor German.

Venus in Furs has caught his soul in the red snares of hair. He will paint her, and go mad.

* * * * *

It is a sunny winter's day. Something that looks like gold trembles on the leaves of the clusters of trees down below in the green level of the meadow. The camelias at the foot of the gallery are glorious in their abundant buds.

Wanda is sitting in the loggia; she is drawing. The German painter stands opposite her with his hands folded as in adoration, and looks at her. No, he rather looks at her face, and is entirely absorbed in it, enraptured.

But she does not see him, neither does she see me, who with the spade in my hand am turning over the flower-bed, solely that I may see her and feel her nearness, which produces an effect on me like poetry, like music.

* * * * *

The painter has gone. It is a hazardous thing to do, but I risk it. I go up to the gallery, quite close, and ask Wanda "Do you love the painter, mistress?"

She looks at me without getting angry, shakes her head, and finally even smiles.

"I feel sorry for him," she replies, "but I do not love him. I love no one. *I used to love you, as ardently, as passionately, as deeply as it was possible for me to love,* but now I don't love even you any more; my heart is a void, dead, and this makes me sad."

"Wanda!" I exclaimed, deeply moved.

"Soon, you too will no longer love me," she continued, "tell me when you have reached that point, and I will give back to you your freedom."

"Then I shall remain your slave, all my life long, for I adore you and shall always adore you," I cried, seized by that fanaticism of love which has repeatedly been so fatal to me.

Wanda looked at me with a curious pleasure. "Consider well what you do," she said. "I have loved you infinitely and have been despotic towards you so that I might fulfil your dream. Something of my old feeling, a sort of real sympathy for you, still trembles in my breast. When that too has gone who knows whether then I shall give you your liberty; whether I shall not then become really cruel, merciless, even brutal toward; whether I shall not take a diabolical pleasure in tormenting and putting on the rack the man who worships me idolatrously, the while I remain indifferent or love someone else; perhaps, I shall enjoy seeing him die of his love for me. Consider this well."

"I have long since considered all that," I replied as in a glow of fever. "I cannot exist, cannot live without you; I shall die if you set me at liberty; let me remain your slave, kill me, but do not drive me away."

"Very well then, be my slave," she replied, "but don't forget that I no longer love you, and your love doesn't mean any more to me than a dog's, and dogs are kicked."

* * * * *

To-day I visited the Venus of Medici.

It was still early, and the little octagonal room in the Tribuna was filled with half-lights like a sanctuary; I stood with folded hands in deep adoration before the silent image of the divinity.

But I did not stand for long.

Not a human soul was in the gallery, not even an Englishman, and I fell down on my knees. I looked up at the lovely slender body, the budding breasts, the virginal and yet voluptuous face, the fragrant curls which seemed to conceal tiny horns on each side of the forehead.

* * * * *

My mistress's bell.

It is noonday. She, however, is still abed with her arms intertwined behind her neck.

"I want to bathe," she says, "and you will attend me. Lock the door!"

I obey.

"Now go downstairs and make sure the door below is also locked."

I descended the winding stairs that lead from her bedroom to the bath; my feet gave way beneath me, and I had to support myself against the iron banister. After having ascertained that the door leading to the Loggia and the garden was locked, I returned. Wanda was now sitting on the bed with loosened hair, wrapped in her green velvet furs. When she made a rapid movement, I noticed that the furs were her only covering. It made me start terribly, I don't know why? I was like one condemned to death, who knows he is on the way to the scaffold, and yet begins to tremble when he sees it.

"Come, Gregor, take me on your arms."

"You mean, mistress?"

"You are to carry me, don't you understand?"

I lifted her up, so that she rested in my arms, while she twined hers around my neck. Slowly, step by step, I went down the stairs with her and her hair beat from time to time against my cheek and her foot sought support against my knee. I trembled under the beautiful burden I was carrying, and every moment it seemed as if I had to break down beneath it.

The bath consisted of a wide, high rotunda, which received a soft quiet light from a red glass cupola above. Two palms extended their broad leaves like a roof over a couch of velvet cushions. From here steps covered with Turkish rugs led to the white marble basin which occupied the center.

"There is a green ribbon on my toilet-table upstairs," said Wanda, as I let her down on the couch, "go get it, and also bring the whip."

I flew upstairs and back again, and kneeling put both in my mistress's hands. She then had me twist her heavy electric hair into a large knot which I fastened with the green ribbon. Then I prepared the bath. I did this very awkwardly because my hands and feet refused to obey me. Again and again I had to look at the beautiful woman lying on the red velvet cushions, and from time to time her wonderful body gleamed here and there beneath the furs. Some magnetic power stronger than my will compelled me to look. I felt that all sensuality and lustfulness lies in that which is half-concealed or intentionally disclosed; and the truth of this I recognized even more acutely, when the basin at last was full, and Wanda threw off the fur- cloak with a single gesture, and stood before me like the goddess in the Tribuna.

At that moment she seemed as sacred and chaste to me in her unveiled beauty, as did the divinity of long ago. I sank down on my knees before her, and devoutly pressed my lips on her foot.

My soul which had been storm-tossed only a little while earlier, suddenly was perfectly calm, and I now felt no element of cruelty in Wanda.

She slowly descended the stairs, and I could watch her with a calmness in which not a single atom of torment or desire was intermingled. I could see her plunge into and rise out of the crystalline water, and the wavelets which she herself raised played about her like tender lovers.

Our nihilistic aesthetician is right when he says: a real apple is more beautiful than a painted one, and a living woman is more beautiful than a Venus of stone.

And when she left the bath, and the silvery drops and the roseate light rippled down her body, I was seized with silent rapture. I wrapped the linen sheets about her, drying her glorious body. The calm bliss remained with me, even now when one foot upon me as upon a footstool, she rested on the cushions in her large velvet cloak. The lithe sables nestled desirously against her cold marble-like body. Her left arm on which she supported herself lay like a sleeping swan in the dark fur of the sleeve, while her left hand played carelessly with the whip.

By chance my look fell on the massive mirror on the wall opposite, and I cried out, for I saw the two of us in its golden frame as in a picture. The picture was so marvellously beautiful, so strange, so imaginative, that I was filled with deep sorrow at the thought that its lines and colors would have to dissolve like mist.

"What is the matter?" asked Wanda.

I pointed to the mirror.

"Ah, that is really beautiful," she exclaimed, "too bad one can't capture the moment and make it permanent."

"And why not?" I asked. "Would not any artist, even the most famous, be proud if you gave him leave to paint you and make you immortal by means of his brush.

"The very thought that this extra-ordinary beauty is to be lost to the world," I continued still watching her enthusiastically, "is horrible--all this glorious facial expression, this mysterious eye with its green fires, this demonic hair, this magnificence of body. The idea fills me with a horror of death, of annihilation. But the hand of an artist shall snatch you from this. You shall not like the rest of us disappear absolutely and forever, without leaving a trace of your having been. Your picture must live, even when you yourself have long fallen to dust; your beauty must triumph beyond death!"

Wanda smiled.

"Too bad, that present-day Italy hasn't a Titian or Raphael," she said, "but, perhaps, love will make amends for genius, who knows; our little German might do?" She pondered.

"Yes, he shall paint you, and I will see to it that the god of love mixes his colors."

* * * * *

The young painter has established his studio in her villa; he is completely in her net. He has just begun a Madonna, a Madonna with red hair and green eyes! Only the idealism of a German would attempt to use this thorough-bred woman as a model for a picture of virginity. The poor fellow really is an almost bigger donkey than I am. Our misfortune is that our Titania has discovered our ass's ears too soon.

* * * * *

Now she laughs derisively at us, and how she laughs! I hear her insolent melodious laughter in his studio, under the open window of which I stand, jealously listening.

* * * * *

"Are you mad, me--ah, it is unbelievable, me as the Mother of God!" she exclaimed and laughed again. "Wait a moment, I will show you another picture of myself, one that I myself have painted, and you shall copy it."

Her head appeared in the window, luminous like a flame under the sunlight.

"Gregor!"

I hurried up the stairs, through the gallery, into the studio.

"Lead him to the bath," Wanda commanded, while she herself hurried away.

A few moments passed and Wanda arrived; dressed in nothing but the sable fur, with the whip in her hand; she descended the stairs and stretched out on the velvet cushions as on the former occasion. I lay at her feet and she placed one of her feet upon me; her right hand played with the whip. "Look at me," she said, "with your deep, fanatical look, that's it."

The painter had turned terribly pale. He devoured the scene with his beautiful dreamy blue eyes; his lips opened, but he remained dumb.

"Well, how do you like the picture?"

"Yes, that is how I want to paint you," said the German, but it was really not a spoken language; it was the eloquent moaning, the weeping of a sick soul, a soul sick unto death.

* * * * *

The charcoal outline of the painting is done; the heads and flesh parts are painted in. Her diabolical face is already becoming visible under a few bold strokes, life flashes in her green eyes.

Wanda stands in front of the canvas with her arms crossed over her breast.

"This picture, like many of those of the Venetian school, is simultaneously to represent a portrait and to tell a story," explained the painter, who again had become pale as death.

"And what will you call it?" she asked, "but what is the matter with you, are you ill?"

"I am afraid--" he answered with a consuming look fixed on the beautiful woman in furs, "but let us talk of the picture."

"Yes, let us talk about the picture."

"I imagine the goddess of love as having descended from Mount Olympus for the sake of some mortal man. And always cold in this modern world of ours, she seeks to keep her sublime body warm in a large heavy fur and her feet in the lap of her lover. I imagine the favorite of a beautiful despot, who whips her slave, when she is tired of kissing him, and the more she treads him underfoot, the more insanely he loves her. And so I shall call the picture: *Venus in Furs*."

* * * * *

The painter paints slowly, but his passion grows more and more rapidly. I am afraid he will end up by committing suicide. She plays with him and propounds riddles to him which he cannot solve, and he feels his blood congealing in the process, but it amuses her.

During the sitting she nibbles at candies, and rolls the paper- wrappers into little pellets with which she bombards him.

"I am glad you are in such good humor," said the painter, "but your face has lost the expression which I need for my picture."

"The expression which you need for your picture," she replied, smiling. "Wait a moment."

She rose, and dealt me a blow with the whip. The painter looked at her with stupefaction, and a child-like surprise showed on his face, mingled with disgust and admiration.

While whipping me, Wanda's face acquired more and more of the cruel, contemptuous character, which so haunts and intoxicates me.

"Is this the expression you need for your picture?" she exclaimed. The painter lowered his look in confusion before the cold ray of her eye.

"It is the expression--" he stammered, "but I can't paint now--"

"What?" said Wanda, scornfully, "perhaps I can help you?"

"Yes--" cried the German, as if taken with madness, "whip me too."

"Oh! With pleasure," she replied, shrugging her shoulders, "but if I am to whip you I want to do it in sober earnest."

"Whip me to death," cried the painter.

"Will you let me tie you?" she asked, smiling.

"Yes--" he moaned--

Wanda left the room for a moment, and returned with ropes.

"Well--are you still brave enough to put yourself into the power of Venus in Furs, the beautiful despot, for better or worse?" she began ironically.

"Yes, tie me," the painter replied dully. Wanda tied his hands on his back and drew a rope through his arms and a second one around his body, and fettered him to the cross-bars of the window. Then she rolled back the fur, seized the whip, and stepped in front of him.

The scene had a grim attraction for me, which I cannot describe. I felt my heart beat, when, with a smile, she drew back her arm for the first blow, and the whip hissed through the air. He winced slightly under the blow. Then she let blow after blow rain upon him, with her mouth half-opened and her teeth flashing between her red lips, until he finally seemed to ask for mercy with his piteous, blue eyes. It was indescribable.

* * * * *

She is sitting for him now, alone. He is working on her head.

She has posted me in the adjoining room behind a heavy curtain, where I can't be seen, but can see everything.

What does she intend now?

Is she afraid of him? She has driven him insane enough to be sure, or is she hatching a new torment for me? My knees tremble.

They are talking. He has lowered his voice so that I cannot understand a word, and she replies in the same way. What is the meaning of this? Is there an understanding between them?

I suffer frightful torments; my heart seems about to burst.

He kneels down before her, embraces her, and presses his head against her breast, and she--in her heartlessness--laughs--and now I hear her saying aloud:

"Ah! You need another application of the whip."

"Woman! Goddess! Are you without a heart--can't you love," exclaimed the German, "don't you even know, what it means to love, to be consumed with desire and passion, can't you even imagine what I suffer? Have you no pity for me?"

"No!" she replied proudly and mockingly, "but I have the whip."

She drew it quickly from the pocket of her fur-coat, and struck him in the face with the handle. He rose, and drew back a couple of paces.

"Now, are you ready to paint again?" she asked indifferently. He did not reply, but again went to the easel and took up his brush and palette.

The painting is marvellously successful. It is a portrait which as far as the likeness goes couldn't be better, and at the same time it seems to have an ideal quality. The colors glow, are supernatural; almost diabolical, I would call them.

The painter has put all his sufferings, his adoration, and all his execration into the picture.

* * * * *

Now he is painting me; we are alone together for several hours every day. To-day he suddenly turned to me with his vibrant voice and said:

"You love this woman?"

"Yes."

"I also love her." His eyes were bathed in tears. He remained silent for a while, and continued painting.

"We have a mountain at home in Germany within which she dwells," he murmured to himself. "She is a demon."

* * * * *

The picture is finished. She insisted on paying him for it, munificently, in the manner of queens.

"Oh, you have already paid me," he said, with a tormented smile, refusing her offer.

Before he left, he secretly opened his portfolio, and let me look inside. I was startled. Her head looked at me as if out of a mirror and seemed actually to be alive.

"I shall take it along," he said, "it is mine; she can't take it away from me. I have earned it with my heart's blood."

* * * * *

"I am really rather sorry for the poor painter," she said to me to- day, "it is absurd to be as virtuous as I am. Don't you think so too?"

I did not dare to reply to her.

"Oh, I forgot that I am talking with a slave; I need some fresh air, I want to be diverted, I want to forget.

"The carriage, quick!"

Her new dress is extravagant: Russian half-boots of violet-blue velvet trimmed with ermine, and a skirt of the same material, decorated with narrow stripes and rosettes of furs. Above it is an appropriate, close-fitting jacket, also richly trimmed and lined with ermine. The headdress is a tall cap of ermine of the style of Catherine the Second, with a small aigrette, held in place by a diamond-agraffe; her red hair falls loose down her back. She ascends on the driver's seat, and holds the reins herself; I take my seat behind. How she lashes on the horses! The carriage flies along like mad.

Apparently it is her intention to attract attention to-day, to make conquests, and she succeeds completely. She is the lioness of the Cascine. People nod to her from carriages; on the footpath people gather in groups to discuss her. She pays no attention to anyone, except now and then acknowledging the greetings of elderly gentlemen with a slight nod.

Suddenly a young man on a lithe black horse dashes up at full speed. As soon as he sees Wanda, he stops his horse and makes it walk. When he is quite close, he stops entirely and lets her pass. And she too sees him--the lioness, the lion. Their eyes meet. She madly drives past him, but she cannot tear herself free from the magic power of his look, and she turns her head after him.

My heart stops when I see the half-surprised, half-enraptured look with which she devours him, but he is worthy of it.

For he is, indeed, a magnificent specimen of man, No, rather, he is a man whose like I have never yet seen among the living. He is in the Belvedere, graven in marble, with the same slender, yet steely musculature, with the same face and the same waving curls. What makes him particularly beautiful is that he is beardless. If his hips were less narrow, one might take him for a woman in disguise. The curious expression about the mouth, the lion's lip which slightly discloses the teeth beneath, lends a flashing tinge of cruelty to the beautiful face--

Apollo flaying Marsyas.

He wears high black boots, closely fitting breeches of white leather, short fur coat of black cloth, of the kind worn by Italian cavalry officers, trimmed with astrakhan and many rich loops; on his black locks is a red fez.

I now understand the masculine Eros, and I marvel at Socrates for having remained virtuous in view of an Alcibiades like this.

* * * * *

I have never seen my lioness so excited. Her cheeks flamed when she left from the carriage at her villa. She hurried upstairs, and with an imperious gesture ordered me to follow.

Walking up and down her room with long strides, she began to talk so rapidly, that I was frightened.

"You are to find out who the man in the Cascine was, immediately--

"Oh, what a man! Did you see him? What do you think of him? Tell me."

"The man is beautiful," I replied dully.

"He is so beautiful," she paused, supporting herself on the arm of a chair, "that he has taken my breath away."

"I can understand the impression he has made on you," I replied, my imagination carrying me away in a mad whirl. "I am quite lost in admiration myself, and I can imagine--"

"You may imagine," she laughed aloud, "that this man is my lover, and that he will apply the lash to you, and that you will enjoy being punished by him.

"But now go, go."

* * * * *

Before evening fell, I had the desired information.

Wanda was still fully dressed when I returned. She reclined on the ottoman, her face buried in her hands, her hair in a wild tangle, like the red mane of a lioness.

"What is his name?" she asked, uncanny calm.

"Alexis Papadopolis."

"A Greek, then,"

I nodded.

"He is very young?"

"Scarcely older than you. They say he was educated in Paris, and that he is an atheist. He fought against the Turks in Candia, and is said to have distinguished himself there no less by his race-hatred and cruelty, than by his bravery."

"All in all, then, a man," she cried with sparkling eyes.

"At present he is living in Florence," I continued, "he is said to be tremendously rich--"

"I didn't ask you about that," she interrupted quickly and sharply. "The man is dangerous. Aren't you afraid of him? I am afraid of him. Has he a wife?"

"No."

"A mistress?"

"No."

"What theaters does he attend?"

"To-night he will be at the Nicolini Theater, where Virginia Marini and Salvini are acting; they are the greatest living artists in Italy, perhaps in Europe.

"See that you get a box--and be quick about it!" she commanded.

"But, mistress--"

"Do you want a taste of the whip?"

* * * * *

"You can wait down in the lobby," she said when I had placed the opera-glasses and the programme on the edge of her box and adjusted the footstool.

I am standing there and had to lean against the wall for support so as not to fall down with envy and rage--no, rage isn't the right word; it was a mortal fear.

I saw her in her box dressed in blue moire, with a huge ermine cloak about her bare shoulders; he sat opposite. I saw them devour each other with their eyes. For both of them the stage, Goldoni's *Pamela,* Salvini, Marini, the public, even the entire world, were non-existant to-night. And I--what was I at that moment?--

* * * * *

To-day she is attending the ball at the Greek ambassador's. Does she know, that she will meet him there?

At any rate she dressed, as if she did. A heavy sea-green silk dress plastically encloses her divine form, leaving the bust and arms bare. In her hair, which is done into a single flaming knot, a white water- lily blossoms; from it the leaves of reeds interwoven with a few loose strands fall down toward her neck. There no longer is any trace of agitation or trembling feverishness in her being. She is calm, so calm, that I feel my blood congealing and my heart growing cold under her glance. Slowly, with a weary, indolent majesty, she ascends the marble staircase, lets her precious wrap slide off, and listlessly enters the hall, where the smoke of a hundred candles has formed a silvery mist.

For a few moments my eyes follow her in a daze, then I pick up her furs, which without my being aware, had slipped from my hands. They are still warm from her shoulders.

I kiss the spot, and my eyes fill with tears.

* * * * *

He has arrived.

In his black velvet coat extravagantly trimmed with sable, he is a beautiful, haughty despot who plays with the lives and souls of men. He stands in the ante-room, looking around proudly, and his eyes rest on me for an uncomfortably long time.

Under his icy glance I am again seized by a mortal fear. I have a presentiment that this man can enchain her, captivate her, subjugate her, and I feel inferior in contrast with his savage masculinity; I am filled with envy, with jealousy.

I feel that I am a queer weakly creature of brains, merely! And what is most humiliating, I want to hate him, but I can't. Why is that among all the host of servants he has chosen me.

With an inimitably aristocratic nod of the head he calls me over to him, and I--I obey his call--against my own will.

"Take my furs," he quickly commands.

My entire body trembles with resentment, but I obey, abjectly like a slave.

* * * * *

All night long I waited in the ante-room, raving as in a fever. Strange images hovered past my inner eye. I saw their meeting--their long exchange of looks. I saw her float through the hall in his arms, drunken, lying with half-closed lids against his breast. I saw him in the holy of holies of love, lying on the ottoman, not as slave, but as master, and she at his feet. On my knees I served them, the tea-tray faltering in my hands, and I saw him reach for the whip. But now the servants are talking about him.

He is a man who is like a woman; he knows that he is beautiful, and he acts accordingly. He changes his clothes four or five times a day, like a vain courtesan.

In Paris he appeared first in woman's dress, and the men assailed him with love-letters. An Italian singer, famous equally for his art and his passionate intensity, even invaded his home, and lying on his knees before him threatened to commit suicide if he wouldn't be his.

"I am sorry," he replied, smiling, "I should like to do you the favor, but you will have to carry out your threat, for I am a man."

* * * * *

The drawing-room has already thinned out to a marked degree, but she apparently has no thought of leaving.

Morning is already peering through the blinds.

At last I hear the rustling of her heavy gown which flows along behind her like green waves. She advances step by step, engaged in conversation with him.

I hardly exist for her any longer; she doesn't even trouble to give me an order.

"The cloak for madame," he commands. He, of course, doesn't think of looking after her himself.

While I put her furs about her, he stands to one side with his arms crossed. While I am on my knees putting on her fur over-shoes, she lightly supports herself with her hand on his shoulder. She asks:

"And what about the lioness?"

"When the lion whom she has chosen and with whom she lives is attacked by another," the Greek went on with his narrative, "the lioness quietly lies down and watches the battle. Even if her mate is worsted she does not go to his aid. She looks on indifferently as he bleeds to death under his opponent's claws, and follows the victor, the stronger--that is the female's nature."

At this moment my lioness looked quickly and curiously at me.

It made me shudder, though I didn't know why--and the red dawn immerses me and her and him in blood.

* * * * *

She did not go to bed, but merely threw off her ball-dress and undid her hair; then she ordered me to build a fire, and she sat by the fire-place, and stared into the flames.

"Do you need me any longer, mistress?" I asked, my voice failed me at the last word.

Wanda shook her head.

I left the room, passed through the gallery, and sat down on one of the steps, leading from there down into the garden. A gentle north wind brought a fresh, damp coolness from the Arno, the green hills extended into the distance in a rosy mist, a golden haze hovered over the city, over the round cupola of the Duomo.

A few stars still tremble in the pale-blue sky.

I tore open my coat, and pressed my burning forehead against the marble. Everything that had happened so far seemed to me a mere child's play; but now things were beginning to be serious, terribly serious.

I anticipated a catastrophe, I visualized it, I could lay hold of it with my hands, but I lacked the courage to meet it. My strength was broken. And if I am honest with myself, neither the pains and sufferings that threatened me, not the humiliations that impended, were the thing that frightened me.

I merely felt a fear, the fear of losing her whom I loved with a sort of fanatical devotion; but it was so overwhelming, so crushing that I suddenly began to sob like a child.

* * * * *

During the day she remained locked in her room, and had the negress attend her. When the evening star rose glowing in the blue sky, I saw her pass through the garden, and, carefully following her at a distance, watched her enter the shrine of Venus. I stealthily followed and peered through the chink in the door.

She stood before the divine image of the goddess, her hands folded as in prayer, and the sacred light of the star of love casts its blue rays over her.

* * * * *

On my couch at night the fear of losing her and despair took such powerful hold of me that they made a hero and a libertine of me. I lighted the little red oil-lamp which hung in the corridor beneath a saint's image, and entered her bedroom, covering the light with one hand.

The lioness had been hunted and driven until she was exhausted. She had fallen asleep among her pillows, lying on her back, her hands clenched, breathing heavily. A dream seemed to oppress her. I slowly withdrew my hand, and let the red light fall full on her wonderful face.

But she did not awaken.

I gently set the lamp on the floor, sank down beside Wanda's bed, and rested my head on her soft, glowing arm.

She moved slightly, but even now did not awaken. I do not know how long I lay thus in the middle of the night, turned as into a stone by horrible torments.

Finally a severe trembling seized me, and I was able to cry. My tears flowed over her arm. She quivered several times and finally sat up; she brushed her hand across her eyes, and looked at me.

"Severin," she exclaimed, more frightened than angry.

I was unable to reply.

"Severin," she continued softly, "what is the matter? Are you ill?"

Her voice sounded so sympathetic, so kind, so full of love, that it clutched my breast like red-hot tongs and I began to sob aloud.

"Severin," she began anew. "My poor unhappy friend." Her hand gently stroked my hair. "I am sorry, very sorry for you; but I can't help you; with the best intention in the world I know of nothing that would cure you."

"Oh, Wanda, must it be?" I moaned in my agony.

"What, Severin? What are you talking about?"

"Don't you love me any more?" I continued. "Haven't you even a little bit of pity for me? Has the beautiful stranger taken complete possession of you?"

"I cannot lie," she replied softly after a short pause. "He has made an impression on me which I haven't yet been able to analyse, further than that I suffer and tremble beneath it. It is an impression of the sort I have met with in the works of poets or on the stage, but I always thought it was a figment of the imagination. Oh, he is a man like a lion, strong and beautiful and yet gentle, not brutal like the men of our northern world. I am sorry for you, Severin, I am; but I must possess him. What am I saying? I must give myself to him, if he will have me."

"Consider your reputation, Wanda, which so far has remained spotless," I exclaimed, "even if I no longer mean anything to you."

"I am considering it," she replied, "I intend to be strong, as long as it is possible, I want--" she buried her head shyly in the pillows --"I want to become his wife--if he will have me."

"Wanda," I cried, seized again by that mortal fear, which always robs me of my breath, makes me lose possession of myself, "you want to be his wife, belong to him for always. Oh! Do not drive me away! He does not love you--"

"Who says that?" she exclaimed, flaring up.

"He does not love you," I went on passionately, "but I love you, I adore you, I am your slave, I let you tread me underfoot, I want to carry you on my arms through life."

"Who says that he doesn't love me?" she interrupted vehemently.

"Oh! be mine," I replied, "be mine! I cannot exist, cannot live without you. Have mercy on me, Wanda, have mercy!"

She looked at me again, and her face had her cold heartless expression, her evil smile.

"You say he doesn't love me," she said, scornfully. "Very well then, get what consolation you can out of it."

With this she turned over on the other side, and contemptuously showed me her back.

"Good God, are you a woman without flesh or blood, haven't you a heart as well as I!" I cried, while my breast heaved convulsively.

"You know what I am," she replied, coldly. "I am a woman of stone, *Venus in Furs*, your ideal, kneel down, and pray to me."

"Wanda!" I implored, "mercy!"

She began to laugh. I buried my face in her pillows. Pain had loosened the floodgates of my tears and I let them flow.

For a long time silence reigned, then Wanda slowly raised herself.

"You bore me," she began.

"Wanda!"

"I am tired, let me go to sleep."

"Mercy," I implored. "Do not drive me away. No man, no one, will love you as I do."

"Let me go to sleep,"--she turned her back to me again.

I leaped up, and snatched the poinard, which hung beside her bed, from its sheath, and placed its point against my breast.

"I shall kill myself here before your eyes," I murmured dully.

"Do what you please," Wanda replied with complete indifference. "But let me go to sleep." She yawned aloud. "I am very sleepy."

For a moment I stood as if petrified. Then I began to laugh and cry at the same time. Finally I placed the poinard in my belt, and again fell on my knees before her.

"Wanda, listen to me, only for a few moments," I begged.

"I want to go to sleep! Don't you hear!" she cried, leaping angrily out of bed and pushing me away with her foot. "You forget that I am your mistress?" When I didn't budge, she seized the whip and struck me. I rose; she struck me again--this time right in the face.

"Wretch, slave!"

With clenched fist held heavenward, I left her bedroom with a sudden resolve. She tossed the whip aside, and broke out into clear laughter. I can imagine that my theatrical attitude must have been very droll.

* * * * *

I have determined to set myself free from this heartless woman, who has treated me so cruelly, and is now about to break faith and betray me, as a reward for all my slavish devotion, for everything I have suffered from her. I packed my few belongings into a bundle, and then wrote her as follows:

"Dear Madam,--

I have loved you even to madness, I have given myself to you as no man ever has given himself to a woman. You have abused my most sacred emotions, and played an impudent, frivolous game with me. However, as long as you were merely cruel and merciless, it was still possible for me to love you. Now you are about to become *cheap*. I am no longer the slave

whom you can kick about and whip. You yourself have set me free, and I
am leaving a woman I can only hate and despise.

Severin Kusiemski."

I handed these lines to the negress, and hastened away as fast as I could go.
I arrived at the railway-station all out of breath. Suddenly I felt a sharp pain
in my heart and stopped. I began to weep. It is humiliating that I want to
flee and I can't. I turn back-- whither?--to her, whom I abhor, and yet, at the
same time, adore.

Again I pause. I cannot go back. I dare not.

But how am I to leave Florence. I remember that I haven't any money, not a
penny. Very well then, on foot; it is better to be an honest beggar than to
eat the bread of a courtesan.

But still I can't leave.

She has my pledge, my word of honor. I have to return. Perhaps she will
release me.

After a few rapid strides, I stop again.

She has my word of honor and my bond, that I shall remain her slave as
long as she desires, until she herself gives me my freedom. But I might kill
myself.

I go through the Cascine down to the Arno, where its yellow waters plash
monotonously about a couple of stray willows. There I sit, and cast up my
final accounts with existence. I let my entire life pass before me in review.
On the whole, it is rather a wretched affair--a few joys, an endless number
of indifferent and worthless things, and between these an abundant harvest
of pains, miseries, fears, disappointments, shipwrecked hopes, afflictions,
sorrow and grief.

I thought of my mother, whom I loved so deeply and whom I had to watch waste away beneath a horrible disease; of my brother, who full of the promise of joy and happiness died in the flower of youth, without even having put his lips to the cup of life. I thought of my dead nurse, my childhood playmates, the friends that had striven and studied with me; of all those, covered by the cold, dead, indifferent earth. I thought of my turtle-dove, who not infrequently made his cooing bows to me, instead of to his mate.--All have returned, dust unto dust.

I laughed aloud, and slid down into the water, but at the same moment I caught hold of one of the willow-branches, hanging above the yellow waves. As in a vision, I see the woman who has caused all my misery. She hovers above the level of the water, luminous in the sunlight as though she were transparent, with red flames about her head and neck. She turns her face toward me and smiles.

* * * * *

I am back again, dripping, wet through, glowing with shame and fever. The negress has delivered my letter; I am judged, lost, in the power of a heartless, affronted woman.

Well, let her kill me. I am unable to do it myself, and yet I have no wish to go on living.

As I walk around the house, she is standing in the gallery, leaning over the railing. Her face is full in the light of the sun, and her green eyes sparkle.

"Still alive?" she asked, without moving. I stood silent, with bowed head.

"Give me back my poinard," she continued. "It is of no use to you. You haven't even the courage to take your own life."

"I have lost it," I replied, trembling, shaken by chills.

She looked me over with a proud, scornful glance.

"I suppose you lost it in the Arno?" She shrugged her shoulders. "No matter. Well, and why didn't you leave?"

I mumbled something which neither she nor I myself could understand.

"Oh! you haven't any money," she cried. "Here!" With an indescribably disdainful gesture she tossed me her purse.

I did not pick it up.

Both of us were silent for some time.

"You don't want to leave then?"

"I can't."

* * * * *

Wanda drives in the Cascine without me, and goes to the theater without me; she receives company, and the negress serves her. No one asks after me. I stray about the garden, irresolutely, like an animal that has lost its master.

Lying among the bushes, I watch a couple of sparrows, fighting over a seed.

Suddenly I hear the swish of a woman's dress.

Wanda approaches in a gown of dark silk, modestly closed up to the neck; the Greek is with her. They are in an eager discussion, but I cannot as yet understand a word of what they are saying. He stamps his foot so that the gravel scatters about in all directions, and he lashes the air with his riding whip. Wanda startles.

Is she afraid that he will strike her?

Have they gone that far?

He has left her, she calls him; he does not hear her, does not want to hear her.

Wanda sadly lowers her head, and then sits down on the nearest stone-bench. She sits for a long time, lost in thought. I watch her with a sort of malevolent pleasure, finally I pull myself together by sheer force of will, and ironically step before her. She startles, and trembles all over.

"I come to wish you happiness," I said, bowing, "I see, my dear lady, too, has found a master."

"Yes, thank God!" she exclaimed, "not a new slave, I have had enough of them. A master! Woman needs a master, and she adores him."

"You adore him, Wanda?" I cried, "this brutal person--"

"Yes, I love him, as I have never loved any one else."

"Wanda!" I clenched my fists, but tears already filled my eyes, and I was seized by the delirium of passion, as by a sweet madness. "Very well, take him as your husband, let him be your master, but I want to remain your slave, as long as I live."

"You want to remain my slave, even then?" she said, "that would be interesting, but I am afraid he wouldn't permit it."

"He?"

"Yes, he is already jealous of you," she exclaimed, "he, of you! He demanded that I dismiss you immediately, and when I told him who you were--"

"You told him--" I repeated, thunderstruck.

"I told him everything," she replied, "our whole story, all your queerness, everything--and he, instead of being amused, grew angry, and stamped his foot."

"And threatened to strike you?"

Wanda looked to the ground, and remained silent.

"Yes, indeed," I said with mocking bitterness, "you are afraid of him, Wanda!" I threw myself down at her feet, and in my agitation embraced her knees. "I don't want anything of you, except to be your slave, to be always near you! I will be your dog-"

"Do you know, you bore me?" said Wanda, indifferently.

I leaped up. Everything within me was seething.

"You are now no longer cruel, but cheap," I said, clearly and distinctly, accentuating every word.

"You have already written that in your letter," Wanda replied, with a proud shrug of the shoulders. "A man of brains should never repeat himself."

"The way you are treating me," I broke out, "what would you call it?"

"I might punish you," she replied ironically, "but I prefer this time to reply with reasons instead of lashes. You have no right to accuse me. Haven't I always been honest with you? Haven't I warned you more than once? Didn't I love you with all my heart, even passionately, and did I conceal the fact from you, that it was dangerous to give yourself into my power, to abase yourself before me, and that I want to be dominated? But you wished to be my plaything, my slave! You found the highest pleasure in feeling the foot, the whip of an arrogant, cruel woman. What do you want now?

"Dangerous potentialities were slumbering in me, but you were the first to awaken them. If I now take pleasure in torturing you, abusing you, it is your fault; you have made of me what I now am, and now you are even unmanly, weak, and miserable enough to accuse me."

"Yes, I am guilty," I said, "but haven't I suffered because of it? Let us put an end now to the cruel game."

"That is my wish, too," she replied with a curious deceitful look.

"Wanda!" I exclaimed violently, "don't drive me to extremes; you see that I am a man again."

"A fire of straw," she replied, "which makes a lot of stir for a moment, and goes out as quickly as it flared up. You imagine you can intimidate me, and you only make yourself ridiculous. Had you been the man I first thought you were, serious, reserved, stern, I would have loved you faithfully, and become your wife. Woman demands that she can look up to a man, but one like you who voluntarily places his neck under her foot, she uses as a welcome plaything, only to toss it aside when she is tired of it."

"Try to toss me aside," I said, jeeringly. "Some toys are dangerous."

"Don't challenge me," exclaimed Wanda. Her eyes began to flash, and a flush entered her cheeks.

"If you won't be mine now," I continued, with a voice stifled with rage, "no one else shall possess you either."

"What play is this from?" she mocked, seizing me by the breast. She was pale with anger at this moment. "Don't challenge me," she continued, "I am not cruel, but I don't know whether I may not become so and whether then there will be any bounds."

"What worse can you do, than to make your lover, your husband?" I exclaimed, more and more enraged.

"I might make you *his* slave," she replied quickly, "are you not in my power? Haven't I the agreement? But, of course, you will merely take pleasure in it, if I have you bound, and say to him.

"Do with him what you please."

"Woman, are you mad!" I cried.

"I am entirely rational," she said, calmly. "I warn you for the last time. Don't offer any resistance, one who has gone as far as I have gone might easily go still further. I feel a sort of hatred for you, and would find a real joy in seeing him beat you to death; I am still restraining myself, but--"

Scarcely master of myself any longer, I seized her by the wrist and forced her to the ground, so that she lay on her knees before me.

"Severin!" she cried. Rage and terror were painted on her face.

"I shall kill you if you marry him," I threatened; the words came hoarsely and dully from my breast. "You are mine, I won't let you go, I love you too much." Then I clutched her and pressed her close to me; my right hand involuntarily seized the dagger which I still had in my belt.

Wanda fixed a large, calm, incomprehensible look on me.

"I like you that way," she said, carelessly. "Now you are a man, and at this moment I know I still love you."

"Wanda," I wept with rapture, and bent down over her, covering her dear face with kisses, and she, suddenly breaking into a loud gay laugh, said, "Have you finished with your ideal now, are you satisfied with me?"

"You mean?" I stammered, "that you weren't serious?"

"I am very serious," she gaily continued. "I love you, only you, and you--you foolish, little man, didn't know that everything was only make-believe and play-acting. How hard it often was for me to strike you with the whip, when I would have rather taken your head and covered it with kisses. But now we are through with that, aren't we? I have played my cruel role better than you expected, and now you will be satisfied with my being a good, little wife who isn't altogether unattractive. Isn't that so? We will live like rational people--"

"You will marry me!" I cried, overflowing with happiness.

"Yes--marry you--you dear, darling man," whispered Wanda, kissing my hands.

I drew her up to my breast.

"Now, you are no longer Gregor, my slave," said she, "but Severin, the dear man I love--"

"And he--you don't love him?" I asked in agitation.

"How could you imagine my loving a man of his brutal type? You were blind to everything, I was really afraid for you."

"I almost killed myself for your sake."

"Really?" she cried, "ah, I still tremble at the thought, that you were already in the Arno."

"But you saved me," I replied, tenderly. "You hovered over the waters and smiled, and your smile called me back to life."

* * * *

I have a curious feeling when I now hold her in my arms and she lies silently against my breast and lets me kiss her and smiles. I feel like one who has suddenly awakened out of a feverish delirium, or like a shipwrecked man who has for many days battled with waves that momentarily threatened to devour him and finally has found a safe shore.

* * * * *

"I hate this Florence, where you have been so unhappy," she declared, as I was saying good-night to her. "I want to leave immediately, tomorrow, you will be good enough to write a couple of letters for me, and, while you are doing that, I will drive to the city to pay my farewell visits. Is that satisfactory to you?"

"Of course, you dear, sweet, beautiful woman."

* * * * *

Early in the morning she knocked at my door to ask how I had slept. Her tenderness is positively wonderful. I should never have believed that she could be so tender.

* * * * *

She has now been gone for over four hours. I have long since finished the letters, and am now sitting in the gallery, looking down the street to see whether I cannot discover her carriage in the distance. I am a little worried about her, and yet I know there is no reason under heaven why I should doubt or fear. However, a feeling of oppression weighs me down, and I cannot rid myself of it. It is probably the sufferings of the past days, which still cast their shadows into my soul.

* * * * *

She is back, radiant with happiness and contentment.

"Well, has everything gone as you wished?" I asked tenderly, kissing her hand.

"Yes, dear heart," she replied, "and we shall leave to-night. Help me pack my trunks."

* * * * *

Toward evening she asked me to go to the post-office and mail her letters myself. I took her carriage, and was back within an hour.

"Mistress has asked for you," said the negress, with a grin, as I ascended the wide marble stairs.

"Has anyone been here?"

"No one," she replied, crouching down on the steps like a black cat.

I slowly passed through the drawing-room, and then stood before her bedroom door.

Why does my heart beat so? Am I not perfectly happy?

Opening the door softly, I draw back the portiere. Wanda is lying on the ottoman, and does not seem to notice me. How beautiful she looks, in her silver-gray dress, which fits closely, and while displaying in tell-tale fashion her splendid figure, leaves her wonderful bust and arms bare.

Her hair is interwoven with, and held up by a black velvet ribbon. A mighty fire is burning in the fire-place, the hanging lamp casts a reddish glow, and the whole room is as if drowned in blood.

"Wanda," I said at last.

"Oh Severin," she cried out joyously. "I have been impatiently waiting for you." She leaped up, and folded me in her arms. She sat down again on the rich cushions and tried to draw me down to her side, but I softly slid down to her feet and placed my head in her lap.

"Do you know I am very much in love with you to-day?" she whispered, brushing a few stray hairs from my forehead and kissing my eyes.

"How beautiful your eyes are, I have always loved them as the best of you, but to-day they fairly intoxicate me. I am all--" She extended her magnificent limbs and tenderly looked at me from beneath her red lashes.

"And you--you are cold--you hold me like a block of wood; wait, I'll stir you with the fire of love," she said, and again clung fawningly and caressingly to my lips.

"I no longer please you; I suppose I'll have to be cruel to you again, evidently I have been too kind to you to-day. Do you know, you little fool, what I shall do, I shall whip you for a while--"

"But child--"

"I want to."

"Wanda!"

"Come, let me bind you," she continued, and ran gaily through the room. "I want to see you very much in love, do you understand? Here are the ropes. I wonder if I can still do it?"

She began with fettering my feet and then she tied my hands behind my back, pinioning my arms like those of a prisoner.

"So," she said, with gay eagerness. "Can you still move?"

"No."

"Fine--"

She then tied a noose in a stout rope, threw it over my head, and let it slip down as far as the hips. She drew it tight, and bound me to a pillar.

A curious tremor seized me at that moment.

"I have a feeling as if I were about to be executed," I said with a low voice.

"Well, you shall have a thorough punishment to-day," exclaimed Wanda.

"But put on your fur-jacket, please," I said.

"I shall gladly give you that pleasure," she replied. She got her *kazabaika*, and put it on. Then she stood in front of me with her arms folded across her chest, and looked at me out of half-closed eyes.

"Do you remember the story of the ox of Dionysius?" she asked.

"I remember it only vaguely, what about it?"

"A courtier invented a new implement of torture for the Tyrant of Syracuse. It was an iron ox in which those condemned to death were to be shut, and then pushed into a mighty furnace.

"As soon as the iron ox began to get hot, and the condemned person began to cry out in his torment, his wails sounded like the bellowing of an ox.

"Dionysius nodded graciously to the inventor, and to put his invention to an immediate test had him shut up in the iron ox.

"It is a very instructive story.

"It was you who innoculated me with selfishness, pride, and cruelty, and *you shall be their first victim.* I now literally enjoy having a human being that thinks and feels and desires like myself in my power; I love to abuse a man who is stronger in intelligence and body than I, especially a man who loves me.

"Do you still love me?"

"Even to madness," I exclaimed.

"So much the better," she replied, "and so much the more will you enjoy what I am about to do with you now."

"What is the matter with you?" I asked. "I don't understand you, there is a gleam of real cruelty in your eyes to-day, and you are strangely beautiful--completely *Venus in Furs.*"

Without replying Wanda placed her arms around my neck and kissed me. I was again seized by my fanatical passion.

"Where is the whip?" I asked.

Wanda laughed, and withdrew a couple of steps.

"You really insist upon being punished?" she exclaimed, proudly tossing back her head.

"Yes."

Suddenly Wanda's face was completely transformed. It was as if disfigured by rage; for a moment she seemed even ugly to me.

"Very well, then *you* whip him!" she called loudly.

At the same instant the beautiful Greek stuck his head of black curls through the curtains of her four-poster bed. At first I was speechless, petrified. There was a horribly comic element in the situation. I would have laughed aloud, had not my position been at the same time so terribly cruel and humiliating.

It went beyond anything I had imagined. A cold shudder ran down my back, when my rival stepped from the bed in his riding boots, his tight-fitting white breeches, and his short velvet jacket, and I saw his athletic limbs.

"You are indeed cruel," he said, turning to Wanda.

"Only inordinately fond of pleasure," she replied with a wild sort of humor. "Pleasure alone lends value to existence; whoever enjoys does not easily part from life, whoever suffers or is needy meets death like a friend.

"But whoever wants to enjoy must take life gaily in the sense of the ancient world; he dare not hesitate to enjoy at the expense of others; he must never feel pity; he must be ready to harness others to his carriage or his plough as though they were animals. He must know how to make slaves of men who feel and would enjoy as he does, and use them for his service and pleasure without remorse. It is not his affair whether they like it, or whether they go to rack and ruin. He must always remember this, that if they had him in their power, as he has them they would act in exactly the same way, and he would have to pay for their pleasure with his sweat and blood and soul. That was the world of the ancients: pleasure and cruelty, liberty and slavery went hand in hand. People who want to live like the gods of Olympus must of necessity have slaves whom they can toss into their fish- ponds, and gladiators who will do battle, the while they banquet, and they must not mind if by chance a bit of blood bespatters them."

Her words brought back my complete self-possession.

"Unloosen me!" I exclaimed angrily.

"Aren't you my slave, my property?" replied Wanda. "Do you want me to show you the agreement?"

"Untie me!" I threatened, "otherwise--" I tugged at the ropes.

"Can he tear himself free?" she asked. "He has threatened to kill me."

"Be entirely at ease," said the Greek, testing my fetters.

"I shall call for help," I began again.

"No one will hear you," replied Wanda, "and no one will hinder me from abusing your most sacred emotions or playing a frivolous game with you." she continued, repeating with satanic mockery phrases from my letter to her.

"Do you think I am at this moment merely cruel and merciless, or am I also about to become cheap? What? Do you still love me, or do you already hate and despise me? Here is the whip--" She handed it to the Greek who quickly stepped closer.

"Don't you dare!" I exclaimed, trembling with indignation, "I won't permit it--"

"Oh, because I don't wear furs," the Greek replied with an ironical smile, and he took his short sable from the bed.

"You are adorable," exclaimed Wanda, kissing him, and helping him into his furs.

"May I really whip him?" he asked.

"Do with him what you please," replied Wanda.

"Beast!" I exclaimed, utterly revolted.

The Greek fixed his cold tigerish look upon me and tried out the whip. His muscles swelled when he drew back his arms, and made the whip hiss through the air. I was bound like Marsyas while Apollo was getting ready to flay me.

My look wandered about the room and remained fixed on the ceiling, where Samson, lying at Delilah's feet, was about to have his eyes put out by the Philistines. The picture at that moment seemed to me like a symbol, an eternal parable of passion and lust, of the love of man for woman. "Each one of us in the end is a Samson," I thought, "and ultimately for better or worse is betrayed by the woman he loves, whether he wears an ordinary coat or sables."

"Now watch me break him in," said the Greek. He showed his teeth, and his face acquired the blood-thirsty expression, which startled me the first time I saw him.

And he began to apply the lash--so mercilessly, with such frightful force that I quivered under each blow, and began to tremble all over with pain. Tears rolled down over my cheeks. In the meantime Wanda lay on the ottoman in her fur-jacket, supporting herself on her arm; she looked on with cruel curiosity, and was convulsed with laughter.

The sensation of being whipped by a successful rival before the eyes of an adored woman cannot be described. I almost went mad with shame and despair.

What was most humiliating was that at first I felt a certain wild, supersensual stimulation under Apollo's whip and the cruel laughter of my Venus, no matter how horrible my position was. But Apollo whipped on and on, blow after blow, until I forgot all about poetry, and finally gritted my teeth in impotent rage, and cursed my wild dreams, woman, and love.

All of a sudden I saw with horrible clarity whither blind passion and lust have led man, ever since Holofernes and Agamemnon--into a blind alley, into the net of woman's treachery, into misery, slavery, and death.

It was as though I were awakening from a dream.

Blood was already flowing under the whip. I wound like a worm that is trodden on, but he whipped on without mercy, and she continued to laugh without mercy. In the meantime she locked her packed trunk and slipped into her travelling furs, and was still laughing, when she went downstairs on his arm and entered the carriage.

Then everything was silent for a moment.

I listened breathlessly.

The carriage door slammed, the horse began to pull--the rolling of the carriage for a short time--then all was over.

* * * * *

For a moment I thought of taking vengeance, of killing him, but I was bound by the abominable agreement. So nothing was left for me to do except to keep my pledged word and grit my teeth.

* * * * *

My first impulse after this, the most cruel catastrophe of my life, was to seek laborious tasks, dangers, and privations. I wanted to become a soldier and go to Asia or Algiers, but my father was old and ill and wanted me.

So I quietly returned home and for two years helped him bear his burdens, and learned how to look after the estate which I had never done before. To *labor* and to *do my duty* was comforting like a drink of fresh water. Then my father died, and I inherited the estate, but it meant no change.

I had put on my own Spanish boots and went on living just as rationally as if the old man were standing behind me, looking over my shoulder with his large wise eyes.

One day a box arrived, accompanied by a letter. I recognized Wanda's writing.

Curiously moved, I opened it, and read.

"Sir.--

Now that over three years have passed since that night in Florence, I suppose, I may confess to you that I loved you deeply. You yourself, however, stifled my love by your fantastic devotion and your insane passion. From the moment that you became my slave, I knew it would be impossible for you ever to become my husband. However, I found it interesting to have you realize your ideal in my own person, and, while I gloriously amused myself, perhaps, to cure you.

I found the strong man for whom I felt a need, and I was as happy with him as, I suppose, it is possible for any one to be on this funny ball of clay.

But my happiness, like all things mortal, was of short duration. About a year ago he fell in a duel, and since then I have been living in Paris, like an Aspasia--

And you?--Your life surely is not without its sunshine, if you have gained control of your imagination, and those qualities in you have materialized, which at first so attracted me to you--your clarity of intellect, kindness of heart, and, above all else, your--*moral seriousness.*

I hope you have been cured under my whip; the cure was cruel, but radical. In memory of that time and of a woman who loved you passionately, I am sending you the portrait by the poor German.

Venus in Furs."

I had to smile, and as I fell to musing the beautiful woman suddenly stood before me in her velvet jacket trimmed with ermine, with the whip in her hand. And I continued to smile at the woman I had once loved so insanely, at the fur-jacket that had once so entranced me, at the whip, and ended by smiling at myself and saying: The cure was cruel, but radical; but the main point is, I have been cured.

* * * * *

"And the moral of the story?" I said to Severin when I put the manuscript down on the table.

"That I was a donkey," he exclaimed without turning around, for he seemed to be embarrassed. "If only I had beaten her!"

"A curious remedy," I exclaimed, "which might answer with your peasant-women--"

"Oh, they are used to it," he replied eagerly, "but imagine the effect upon one of our delicate, nervous, hysterical ladies--"

"But the moral?"

"That woman, as nature has created her and as man is at present educating her, is his enemy. She can only be his slave or his despot, but *never his companion.* This she can become only when she has the same rights as he, and is his equal in education and work.

"At present we have only the choice of being hammer or anvil, and I was the kind of donkey who let a woman make a slave of him, do you understand?

"The moral of the tale is this: whoever allows himself to be whipped, deserves to be whipped.

"The blows, as you see, have agreed with me; the roseate supersensual mist has dissolved, and no one can ever make me believe again that these 'sacred apes of Benares' [Footnote: One of Schopenhauer's designations for women.] or Plato's rooster [Footnote: Diogenes threw a plucked rooster into Plato's school and exclaimed: "Here you have Plato's human being."] are the image of God."

End of the Project Gutenberg EBook of Venus in Furs by Leopold von Sacher-Masoch Translated by Fernanda Savage

*** END OF THE PROJECT GUTENBERG EBOOK VENUS IN FURS ***

This file should be named vnsfr10.txt or vnsfr10.zip Corrected EDITIONS of our eBooks get a new NUMBER, vnsfr11.txt VERSIONS based on separate sources get new LETTER, vnsfr10a.txt

Produced by Avinash Kothare, Tom Allen, Tiffany Vergon, Charles Aldarondo, Charles Franks and the Online Distributed Proofreading Team.

Project Gutenberg eBooks are often created from several printed editions, all of which are confirmed as Public Domain in the US unless a copyright notice is included. Thus, we usually do not keep eBooks in compliance with any particular paper edition.

We are now trying to release all our eBooks one year in advance of the official release dates, leaving time for better editing. Please be encouraged to tell us about any error or corrections, even years after the official publication date.

Please note neither this listing nor its contents are final til midnight of the last day of the month of any such announcement. The official release date of all Project Gutenberg eBooks is at Midnight, Central Time, of the last day of the stated month. A preliminary version may often be posted for suggestion, comment and editing by those who wish to do so.

Most people start at our Web sites at: http://gutenberg.net or http://promo.net/pg

These Web sites include award-winning information about Project Gutenberg, including how to donate, how to help produce our new eBooks, and how to subscribe to our email newsletter (free!).

Those of you who want to download any eBook before announcement can get to them as follows, and just download by date. This is also a good way to get them instantly upon announcement, as the indexes our cataloguers produce obviously take a while after an announcement goes out in the Project Gutenberg Newsletter.

http://www.ibiblio.org/gutenberg/etext03 or
ftp://ftp.ibiblio.org/pub/docs/books/gutenberg/etext03

Or /etext02, 01, 00, 99, 98, 97, 96, 95, 94, 93, 92, 92, 91 or 90

Just search by the first five letters of the filename you want, as it appears in our Newsletters.

Information about Project Gutenberg

(one page)

We produce about two million dollars for each hour we work. The time it takes us, a rather conservative estimate, is fifty hours to get any eBook selected, entered, proofread, edited, copyright searched and analyzed, the copyright letters written, etc. Our projected audience is one hundred million readers. If the value per text is nominally estimated at one dollar then we produce $2 million dollars per hour in 2002 as we release over 100 new text files per month: 1240 more eBooks in 2001 for a total of 4000+ We are already on our way to trying for 2000 more eBooks in 2002 If they reach just 1-2% of the world's population then the total will reach over half a trillion eBooks given away by year's end.

The Goal of Project Gutenberg is to Give Away 1 Trillion eBooks! This is ten thousand titles each to one hundred million readers, which is only about 4% of the present number of computer users.

Here is the briefest record of our progress (* means estimated):

eBooks Year Month

1 1971 July 10 1991 January 100 1994 January 1000 1997 August 1500 1998 October 2000 1999 December 2500 2000 December 3000 2001 November 4000 2001 October/November 6000 2002 December* 9000

2003 November* 10000 2004 January*

The Project Gutenberg Literary Archive Foundation has been created to secure a future for Project Gutenberg into the next millennium.

We need your donations more than ever!

As of February, 2002, contributions are being solicited from people and organizations in: Alabama, Alaska, Arkansas, Connecticut, Delaware, District of Columbia, Florida, Georgia, Hawaii, Illinois, Indiana, Iowa, Kansas, Kentucky, Louisiana, Maine, Massachusetts, Michigan, Mississippi, Missouri, Montana, Nebraska, Nevada, New Hampshire, New Jersey, New Mexico, New York, North Carolina, Ohio, Oklahoma, Oregon, Pennsylvania, Rhode Island, South Carolina, South Dakota, Tennessee, Texas, Utah, Vermont, Virginia, Washington, West Virginia, Wisconsin, and Wyoming.

We have filed in all 50 states now, but these are the only ones that have responded.

As the requirements for other states are met, additions to this list will be made and fund raising will begin in the additional states. Please feel free to ask to check the status of your state.

In answer to various questions we have received on this:

We are constantly working on finishing the paperwork to legally request donations in all 50 states. If your state is not listed and you would like to know if we have added it since the list you have, just ask.

While we cannot solicit donations from people in states where we are not yet registered, we know of no prohibition against accepting donations from donors in these states who approach us with an offer to donate.

International donations are accepted, but we don't know ANYTHING about how to make them tax-deductible, or even if they CAN be made deductible, and don't have the staff to handle it even if there are ways.

Donations by check or money order may be sent to:

Project Gutenberg Literary Archive Foundation PMB 113 1739 University Ave. Oxford, MS 38655-4109

Contact us if you want to arrange for a wire transfer or payment method other than by check or money order.

The Project Gutenberg Literary Archive Foundation has been approved by the US Internal Revenue Service as a 501(c)(3) organization with EIN [Employee Identification Number] 64-622154. Donations are tax-deductible to the maximum extent permitted by law. As fund-raising requirements for other states are met, additions to this list will be made and fund-raising will begin in the additional states.

We need your donations more than ever!

You can get up to date donation information online at:

http://www.gutenberg.net/donation.html

***

If you can't reach Project Gutenberg, you can always email directly to:

Michael S. Hart <hart@pobox.com>

Prof. Hart will answer or forward your message.

We would prefer to send you information by email.

**

The Legal Small Print

**

(Three Pages)

To create these eBooks, the Project expends considerable efforts to identify, transcribe and proofread public domain works. Despite these efforts, the Project's eBooks and any medium they may be on may contain "Defects". Among other things, Defects may take the form of incomplete, inaccurate or corrupt data, transcription errors, a copyright or other intellectual property infringement, a defective or damaged disk or other eBook medium, a computer virus, or computer codes that damage or cannot be read by your equipment.

LIMITED WARRANTY; DISCLAIMER OF DAMAGES

But for the "Right of Replacement or Refund" described below, [1] Michael Hart and the Foundation (and any other party you may receive this eBook from as a PROJECT GUTENBERG-tm eBook) disclaims all liability to you for damages, costs and expenses, including legal fees, and [2] YOU HAVE NO REMEDIES FOR NEGLIGENCE OR UNDER STRICT LIABILITY, OR FOR BREACH OF WARRANTY OR CONTRACT, INCLUDING BUT NOT LIMITED TO INDIRECT, CONSEQUENTIAL, PUNITIVE OR INCIDENTAL DAMAGES, EVEN IF YOU GIVE NOTICE OF THE POSSIBILITY OF SUCH DAMAGES.

If you discover a Defect in this eBook within 90 days of receiving it, you can receive a refund of the money (if any) you paid for it by sending an explanatory note within that time to the person you received it from. If you received it on a physical medium, you must return it with your note, and such person may choose to alternatively give you a replacement copy. If you received it electronically, such person may choose to alternatively give you a second opportunity to receive it electronically.

THIS EBOOK IS OTHERWISE PROVIDED TO YOU "AS-IS". NO OTHER WARRANTIES OF ANY KIND, EXPRESS OR IMPLIED, ARE MADE TO YOU AS TO THE EBOOK OR ANY MEDIUM IT MAY BE ON, INCLUDING BUT NOT LIMITED TO WARRANTIES OF MERCHANTABILITY OR FITNESS FOR A PARTICULAR PURPOSE.

Some states do not allow disclaimers of implied warranties or the exclusion
or limitation of consequential damages, so the above disclaimers and
exclusions may not apply to you, and you may have other legal rights.

INDEMNITY

You will indemnify and hold Michael Hart, the Foundation, and its trustees
and agents, and any volunteers associated with the production and
distribution of Project Gutenberg-tm texts harmless, from all liability, cost
and expense, including legal fees, that arise directly or indirectly from any
of the following that you do or cause: [1] distribution of this eBook, [2]
alteration, modification, or addition to the eBook, or [3] any Defect.

DISTRIBUTION UNDER "PROJECT GUTENBERG-tm"

You may distribute copies of this eBook electronically, or by disk, book or
any other medium if you either delete this "Small Print!" and all other
references to Project Gutenberg, or:

[1] Only give exact copies of it. Among other things, this requires that you
do not remove, alter or modify the eBook or this "small print!" statement.
You may however, if you wish, distribute this eBook in machine readable
binary, compressed, mark-up, or proprietary form, including any form
resulting from conversion by word processing or hypertext software, but
only so long as *EITHER*:

[*] The eBook, when displayed, is clearly readable, and does *not* contain
characters other than those intended by the author of the work, although
tilde (~), asterisk (*) and underline (i) characters may be used to convey
punctuation intended by the author, and additional characters may be used
to indicate hypertext links; OR

[*] The eBook may be readily converted by the reader at no expense into
plain ASCII, EBCDIC or equivalent form by the program that displays the
eBook (as is the case, for instance, with most word processors); OR

[*] You provide, or agree to also provide on request at no additional cost, fee or expense, a copy of the eBook in its original plain ASCII form (or in EBCDIC or other equivalent proprietary form).

[2] Honor the eBook refund and replacement provisions of this "Small Print!" statement.

[3] Pay a trademark license fee to the Foundation of 20% of the gross profits you derive calculated using the method you already use to calculate your applicable taxes. If you don't derive profits, no royalty is due. Royalties are payable to "Project Gutenberg Literary Archive Foundation" the 60 days following each date you prepare (or were legally required to prepare) your annual (or equivalent periodic) tax return. Please contact us beforehand to let us know your plans and to work out the details.

WHAT IF YOU *WANT* TO SEND MONEY EVEN IF YOU DON'T HAVE TO?

Project Gutenberg is dedicated to increasing the number of public domain and licensed works that can be freely distributed in machine readable form.

The Project gratefully accepts contributions of money, time, public domain materials, or royalty free copyright licenses. Money should be paid to the: "Project Gutenberg Literary Archive Foundation."

If you are interested in contributing scanning equipment or software or other items, please contact Michael Hart at: hart@pobox.com

[Portions of this eBook's header and trailer may be reprinted only when distributed free of all fees. Copyright (C) 2001, 2002 by Michael S. Hart. Project Gutenberg is a TradeMark and may not be used in any sales of Project Gutenberg eBooks or other materials be they hardware or software or any other related product without express permission.]

*END THE SMALL PRINT! FOR PUBLIC DOMAIN EBOOKS*Ver.02/11/02*END*

Venus in Furs
by Leopold von Sacher-Masoch

A free ebook from http://manybooks.net/